AF572284

Feedback Design of Systems with Significant Uncertainty

MECHANICAL ENGINEERING RESEARCH STUDIES

General Editor: **Professor F. J. Bayley,** *University of Sussex, England*

ENGINEERING DYNAMICS AND CONTROL MONOGRAPH SERIES

Series Editor: **Dr. J. B. Roberts,** *University of Sussex, England*

1. Feedback Design of Systems with Significant Uncertainty
   M. J. Ashworth

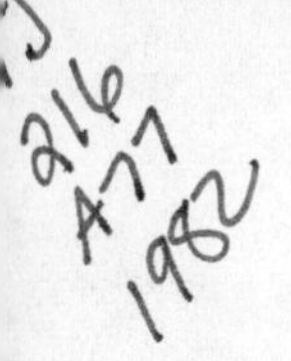

Feedback Design of Systems with Significant Uncertainty

M. J. Ashworth, B.Sc., M.Phil., C.Eng.

Senior Research Fellow, Royal Naval Engineering College, Plymouth, England

RESEARCH STUDIES PRESS
A DIVISION OF JOHN WILEY & SONS LTD.
Chichester · New York · Brisbane · Toronto · Singapore

RESEARCH STUDIES PRESS

Editorial Office:
58B Station Road, Letchworth, Herts. SG6 3BE, England

British Library Cataloguing in Publication Data:

Ashworth, M. J.
Feedback design of systems with significant uncertainty.—(Mechanical engineering research studies).—(Engineering dynamics and control monograph series)
1. Feedback control systems
I. Title II. Series
629.8'3 TJ216
ISBN 0 471 10213 X

Printed in Great Britain

Contents

			Page
EDITORIAL PREFACE			xi
PREFACE			xiii
CHAPTER	1	THE FEEDBACK PRINCIPLE	
	1.1	INTRODUCTION	1
	1.2	THE FEEDBACK CONFIGURATION	2
	1.3	THE BENEFITS OF FEEDBACK	5
	1.4	REDUCTION OF SENSITIVITY TO PLANT VARIATION	5
	1.5	DISTURBANCE REJECTION	7
	1.6	ACHIEVEMENT OF SATISFACTORY DYNAMIC RESPONSE	12
	1.7	NONLINEAR BEHAVIOUR	18
	1.8	SUMMARY	22
CHAPTER	2	CONSIDERATIONS OF DYNAMIC ACCURACY	
	2.1	INTRODUCTION	23
	2.2	PERFORMANCE SPECIFICATIONS AND THE TASK DEFINITION	23
	2.2.1	TIME DOMAIN SPECIFICATIONS	24
	2.2.2	FREQUENCY DOMAIN SPECIFICATIONS	26
	2.2.3	PLANT INPUT EXCITATION	27
	2.3	THE DOMINANT SYSTEM FUNCTION	27
	2.4	SUMMARY	29
CHAPTER	3	LOW ORDER MODELLING	
	3.1	INTRODUCTION	31

			Page
	3.2	REASONS FOR LOW-ORDER MODELLING	32
	3.3	MODELLING FROM THE LOOP FREQUENCY RESPONSE	34
	3.4	MODELLING FROM THE SYSTEM TRANSFER FUNCTION	38
	3.5	MODELLING FROM THE FREQUENCY RESPONSE	43
	3.6	MODELLING USING STANDARD FORMS	47
	3.7	MODELLING FROM THE TIME RESPONSE	51
	3.8	SUMMARY	52
CHAPTER	4	PLANT VARIATION AND SYSTEM SENSITIVITY	
	4.1	INTRODUCTION	55
	4.2	PLANT IGNORANCE	56
	4.3	THE SENSITIVITY FUNCTION	59
	4.4	THE COST OF SENSITIVITY REDUCTION	63
	4.5	SAMPLED-DATA CONSIDERATIONS	65
	4.6	SENSITIVITY ANALYSIS	68
	4.7	COMPONENT TOLERANCING	75
	4.8	SUMMARY	79
CHAPTER	5	LOOP SYNTHESIS AND SENSITIVITY REDUCTION	
	5.1	INTRODUCTION	81
	5.2	ROOT-LOCUS SYNTHESIS	82
	5.3	SYNTHESIS VIA ZERO ASSIGNMENT	84
	5.4	SENSITIVITY REDUCTION VIA THE S-PLANE	88
	5.5	A REMINDER OF NOISE TRANSMISSION	93
	5.6	S-PLANE DESIGN WITH FAR-OFF POLES	94
	5.7	S-PLANE DESIGN WITH VARIABLE PLANT ZEROS	96

Page

5.8 SUMMARY OF S-PLANE DESIGN 96

5.9 FREQUENCY RESPONSE SYNTHESIS 97

5.10 GRAPHICAL SYNTHESIS OF THE LOOP FUNCTION 97

5.11 STABILITY AND THE VARIABLE PLANT 105

5.12 SPECIFICATION OF CLOSED-LOOP TOLERANCES 105

5.13 LOOP SYNTHESIS VIA THE NICHOLS CHART 108

5.14 COMPUTER-AIDED DESIGN 111

5.15 OPTIMIZATION OF THE LOOP FUNCTION 114

5.16 APPLICATIONS TO NONLINEAR SYSTEMS 118

5.17 SUMMARY 124

CHAPTER 6 NOISE DISTURBANCE

6.1 INTRODUCTION 127

6.2 ESTIMATION OF NOISE LEVELS 128

6.3 OPTIMIZATION OF THE SYSTEM PARAMETERS 132

6.4 DETERMINATION OF MSE VIA SIMULATION 135

6.5 THE SYSTEM AS A WIENER FILTER 137

6.6 EXTERNAL NOISE DISTURBANCE 141

6.7 FEEDBACK NOISE 146

6.8 DISTURBANCE CONTROL VIA THE NICHOLS CHART 152

CHAPTER 7 CONSTRAINING THE CONTROL EFFORT

7.1 INTRODUCTION 157

7.2 PERFORMANCE INDICES 158

7.3 AN INDEX TO CONSTRAIN THE CONTROL EFFORT 161

Page

7.4 TRANSFER FUNCTION APPROACH TO OPTIMALITY 162

7.5 THE ROOT-SQUARE-LOCUS SOLUTION 170

7.6 AN OBSERVER-FEEDBACK CONFIGURATION 179

7.7 BODE-DIAGRAM SYNTHESIS 181

7.8 SUMMARY 186

APPENDIX A **ALGORITHMS FOR LOW-ORDER MODELLING**

A1 LEAST-SQUARES COMPLEX CURVE FITTING VIA THE METHOD OF SANATHANAN AND KOERNER 189

A2 COMPUTATION OF CHEN'S CONTINUED-FRACTION MODEL 192

APPENDIX B **LOOP-MAGNITUDE BOUNDARY CIRCLES**

B1 PARAMETERS OF THE BOUNDARY CIRCLES 195

APPENDIX C **CHARACTERISTICS OF SYSTEM FUNCTIONS**

BODE'S THEOREMS 199

CHARACTERISTICS OF THE IDEAL LOOP FUNCTION 207

THE IDEAL LOOP-CHARACTERISTIC OF BODE 210

ALTERNATIVE 'IDEAL' CHARACTERISTICS 214

APPENDIX D **PARSEVAL'S THEOREM**

D1 PARSEVAL'S THEOREM 219

D2 TABULATED INTEGRALS 220

APPENDIX E **EXTREMIZATION AND OPTIMALITY**

E1 EXTREMIZATION OF FUNCTIONS SUBJECT TO CONSTRAINTS 223

E2 EXTREMIZATION OF FUNCTIONALS SUBJECT TO CONSTRAINTS 226

Page

REFERENCES 233

INDEX 243

Editorial Preface

The very rapidly growing literature in the field of Engineering Dynamics and Control is to be found scattered over a wide range of academic and technical journals, conference proceedings and internal reports. This wide distribution often makes it difficult for practising engineers with particular applications to readily locate the information they require on current knowledge. Certain basic information is contained in a variety of text-books but these are often insufficiently specialised to be of direct use to those faced with specific problems.

This volume is the first in a series of monographs. The aim of the series is to identify certain aspects in the general area of Engineering Dynamics and Control where significant research progress has and is being made, and to present coherent and concise surveys of this work which will enable the current 'state of the art' knowledge to be assessed.

The present monograph is concerned with the establishment of rational design procedures for feedback control systems and is founded upon classical linear control theory and frequency domain synthesis. Emphasis is placed upon the design of feedback systems for particular applications which can be realised in terms of hardware. Methods of reducing the effects of noise disturbance are discussed and extension to the multivariable case is suggested. A variety of examples drawn from specific applications are included.

J.B.Roberts

University of Sussex

Preface

Ever since those early days of James Watt and Clerk Maxwell there has been an explosive growth in both the theory and the application of the feedback principle to hardware, biological, economic and management systems stimulated by intense research and development during and following World War II. There are now countless feedback devices operating throughout the world. No one doubts that feedback is a useful concept and that many systems could not function without it. Universality of application has, however, resulted in an unusual phenomenon. Because it is thought to be such a good thing, the properties of feedback have received far less attention than they should. There are two consequences of this ignorance. Firstly, feedback (which costs money) is often so badly used in its applications that the functioning of the closed-loop system is actually inferior to that of the original plant. Secondly, because feedback is not properly exploited, much more sophisticated adaptive systems are designed and manufactured for situations where simpler feedback designs are operationally adequate, cheaper and more reliable.

Such ignorance is surprising since many of the basic ground rules with regard to feedback were enumerated by H.W. Bode in his classic work on the design of radio amplifiers, leading to his famous theorems and to his concept of the ideal loop characteristic. However, despite the extravagance of the mathematical terminology used in arduously difficult problems with which some researchers grapple, a number of modern approaches to system design are primarily concerned with the ultimate task of the definition of the system response alone. Few attempts are made to allow in a quantitative manner for the fact that the mathematical model of the hardware to be controlled is only a theoretical representation of a plant which will manifest all the variations of the manufacturing process and which will change according to age, environment and its operational mode throughout its life.

The motivation for this monograph springs from the awareness that there is a real need to establish securely the foundations of the graduate engineer in the subject - for him to be fully conscious of the need for and the implications of feedback control in any given situation. Only then will he be in a position to have a greater appreciation of, and an accumulated background to judge, the pertinence of the many methods of system synthesis which he will encounter as his experience grows. The text is based upon a series of lectures on control system design presented to final-year engineering undergraduates at the Royal Naval Engineering College, Plymouth.

The monograph endeavours to present feedback as the solution to the fundamental problems of automatic control as distinct to its use as a convenient means of signal shaping to achieve satisfactory dynamic response. Then follows, in a classical setting, the important design methods which have been developed over the years insofar as system synthesis is concerned with the significant problems of the discipline, namely the engineer's ignorance of his system and the environment within which it operates. Stemming from the outstanding work of Bode, Truxal and Horowitz, the engineer has been provided with a fundamental basis of theory upon which he can approach with acumen the design of feedback controllers for realistic plants which will almost invariably exhibit a degree of uncertainty, sometimes of massive proportions, and which are also subject to uncertain external influences. The synthesis is carried out in the frequency domain by virtue of its familiarity to designers and its transparency as a design medium, but principally because of the practical usefulness of the results which can be achieved. Indeed, it is the contention of some researchers in the field that the alternative algebraic-oriented attitude towards system synthesis which is so prevalent today has been singularly ineffective in overcoming real problems and has contributed little of immediate application to the question of sensitivity reduction.

The essential theme of the subject matter is less the achievement of satisfactory dynamic performance than the synthesis of a feedback loop function which will render the hardware plant less troublesome to the operational behaviour of the system and more amenable to the subsequent design stages which will be directed towards dynamic response. The latter area, although considered very briefly in Chapter 2, has been covered more than adequately by Professor D.R.Towill in a companion monograph of this series to which the reader is referred. To meet dynamic response and sensitivity reduction specifications simultaneously, the need to provide two degrees-of-freedom is emphasised. In effect this decouples the two specifications, allowing for individual design procedures. All of the important theorems and fundamental constraints which bear on the loop function and which were first enunciated by Bode and Horowitz have been summarized in Appendix (C) for easy access. A chapter has also been included on low-order modelling since it has been the experience of the author that only by such means can the engineer visualise the salient features of his system, be guided more readily in the selection of the necessary compensation and have before him a clearer objective towards which to aim.

The problem of externally-induced noise and disturbance is introduced and shown to be capable of solution by the very methods used for sensitivity reduction. For a more advanced treatment of noise, particularly where the plant has more than two degrees-of-freedom, the reader is urged to consult the excellent treatise of Horowitz (1963). Feedback noise can impose severe constraints on the attainable loop function and result in power-drive saturation. For this reason, the necessity to constrain the control effort is discussed and which leads quite naturally to the definition of the optimum system configuration via spectral factorization techniques. In keeping with the text, Chang's approach is adopted and which arrives at the optimum system transfer function, the implementation of which is under the total authority of the designer.

Although reference is given within the body of the monograph to the application of the design methods to systems which are categorized as multi-variable, it will be noticed that attention is centred mainly on the single-input/single-output case. To do otherwise would be to move beyond the objectives which have been set. However, again within the philosophy of the text, it is suggested that the most appropriate and useful of the multi-variable design approaches is the frequency-domain Nyquist Array method of Rosenbrock. Here, the plant is pre-compensated to achieve diagonal dominance which implies an 'adequate' reduction of interaction. Once this has been achieved, each loop can be handled independently of one another thus allowing the use of all single-loop techniques.

The author wishes to express his appreciation to all who have assisted in the preparation of this monograph. To the Captain and Dean of the Royal Naval Engineering College, Plymouth, for their support and especially to Commander Roy Savill, Royal Navy, for his continued encouragement and enthusiasm. In particular, the author is indebted to Professor Denis Towill, University of Wales Institute of Science and Technology, for his critical scrutiny of the manuscript, his suggestions for its improvement and as a constant source of inspiration over many years.

CHAPTER 1
The Feedback Principle

1.1. INTRODUCTION

The subject of feedback control is concerned with the automatic control of physical systems and was intended originally to relieve the operator of responsibility for their satisfactory operation and the need for his skill. However, with the advance of the subject, the potential and objectives have become more sophisticated and ambitious. Although feedback devices have been known since the third century B.C. (1), it was the invention of the centrifugal governor by James Watt in 1788 which led to the widespread interest in and the application of the feedback principle wheroby a prescribed relationship is maintained between physical variables by using their difference as an actuating signal to correct for the deviation. The concept of feedback is not associated nowadays with a particular medium as it was in the era of rapid development which occurred during and after World War II, when the servomechanism was in its heyday. Its consequences are to be found in all fields of engineering endeavour and ever increasingly in economic planning, management and the biological sciences. In its broader context, involving both the organic and inorganic together with their interaction, the subject is known by the elevated title of cybernetics, a term first coined by Wiener in 1947.

The principal concern of feedback is the attainment of an objective(s) defined by a set of specifications, by employing a set of equipment (using the term in its most general sense) known collectively as the plant which is activated by a signal(s) derived from the objective and the instantaneous behaviour of the plant. Experience shows that the application of feedback in an indiscriminate manner can lead to numerous problems of design and system performance. When it is considered that if the objective of the design is simply to relate the response of the system to some given demand or, say to speed up the response of the plant, its achievement can often be assured much more economically by pre-filtering than by using feedback, one is naturally led to question the use of feedback at all. Needless to say there is a host of problems of a rather difficult nature for which it is imperative to invoke a feedback configuration about the plant. These topics will be discussed in the remainder of this chapter.

1.2. THE FEEDBACK CONFIGURATION

We consider first the open-loop configuration of figure (1) in which the plant P(s) is acted upon by a signal u(s) obtained by pre-filtering the reference or input signal r(s). Included in the figure are two extraneous inputs, a noise contamination n_i which enters with the reference and a signal n_d shown diagrammatically as entering at the plant input which may consist of noise generated internally, or arise as an external influence or disturbance upon the system.

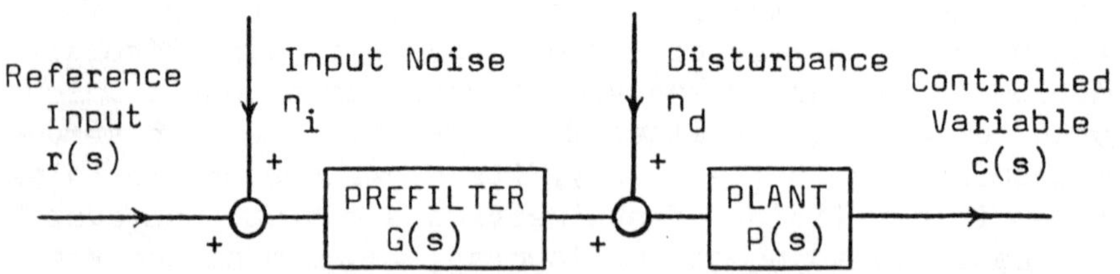

Fig (1): Open-loop control

The prefilter G(s) will have been chosen to 'shape' the system input in such a manner as to enable the plant to provide some desired output response c(s). However, it is apparent that the plant will also respond to the unwanted disturbances n_i and n_d which, being quantities of unknown nature, cannot be cancelled by modification of the system input. Furthermore, since it is possible that the plant may change with time and environmental conditions as well as responding differently to plant inputs which are able to stimulate nonlinear behaviour, the plant response which is achieved may differ widely from that which was intended. We can see therefore that the problems which must be overcome, but which are incapable of solution using the open-loop configuration are as follows:

a. Reduction of the sensitivity of the system to variations in the plant.

b. Reduction of the effect of external noise and disturbance on the plant response.

c. Reduction of the distortion which results from nonlinear operation of the plant.

In addition, there are those plants which exhibit unstable behaviour prior to any compensation. It might be thought that this difficulty could be overcome simply by including terms in G(s) to cancel the unstable terms of the plant. This suggestion is totally impractical since the plant parameters cannot be known precisely for exact cancellation and further, since n_d will always be present to excite the unstable modes.

It should now be obvious that the open-loop structure is unsuited to our purpose and that what is required is a means of modifying the plant input to correct for the deviations of the plant output from the desired output. This can be accomplished simply by introducing a feedback path from output to input as shown in figure (2a). It will be noticed with this configuration that it is impossible to realize independently the overall system transfer function T(s) relating output to input and the loop function L(s) which are defined:

$$T(s) = \frac{L(s)}{1 + L(s)} \quad : \quad L(s) = G(s)P(s) \qquad 1.1$$

For this reason the structure of figure (2a) is said to have one degree-of-freedom. As will become apparent later, it is essential for the development of a useful design method to have independent control over both the fundamental functions T(s) and L(s). This is achieved only by extending the system to have two degrees-of-freedom by adopting the configuration of figure (2b), for which:

$$T(s) = \frac{G(s)P(s)}{1 + L(s)} \quad : \quad L(s) = G(s)P(s)H(s) \qquad 1.2$$

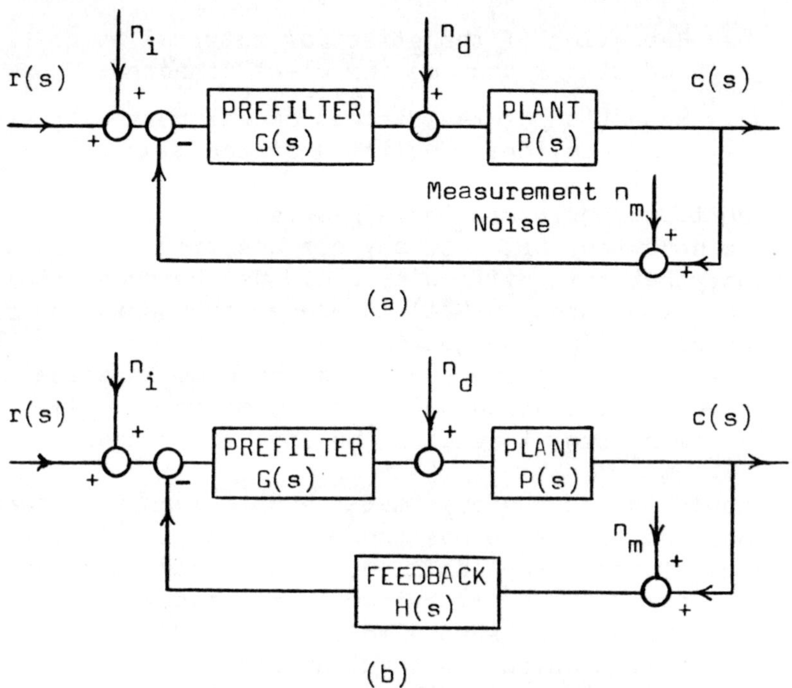

Fig (2): Definition of the single and two degree-of-freedom systems when the plant is accessible only via its input.

It will transpire that the loop function L(s) is of major importance in obtaining the benefits of feedback and is designed independently of the closed-loop transfer function T(s). Having decided upon a suitable loop function, the prefilter will be selected according to our requirements for T(s) thus enabling H(s) to be found by direct substitution.

1.3. THE BENEFITS OF FEEDBACK

It has been intimated without any substantiating proofs that solutions to the problems of control engineering would be forthcoming if some form of feedback was built into the design. The objective of the remainder of this chapter is therefore to examine this claim by posing a number of the difficulties encountered in design so enabling the reader to obtain an impression of just what can be achieved. For the moment, he should view the following examples in a qualitative rather than a quantitative manner since they are used simply as a vehicle to highlight particular problems found with real plants and to illustrate the benefits to be gained when feedback is used.

1.4. REDUCTION OF SENSITIVITY TO PLANT VARIATION

Suppose there exists a plant which can be represented by the block diagram of figure (3a) for which it is known that the only variable parameter K has a nominal value of unity. Let it be assumed that for a variety of reasons it is possible for K to increase in value by up to five times. The objective of the design is simply to compensate the plant in such a manner that the step response is critically damped with an overall system function T(s) given by:

$$T(s) = \frac{1}{s^2 + 2s + 1} \qquad 1.3$$

The transfer function of equation (1.3) can be achieved in an open-loop manner by prefiltering the plant signal as shown in figure (3b) when K assumes its nominal value. Alternatively, the feedback configuration of figure (3c) can be employed where, in addition to the unity feedback of the output c(t), the rate of change of the output is also fed back. Again, with K assuming its nominal value of unity, T(s) is achieved.

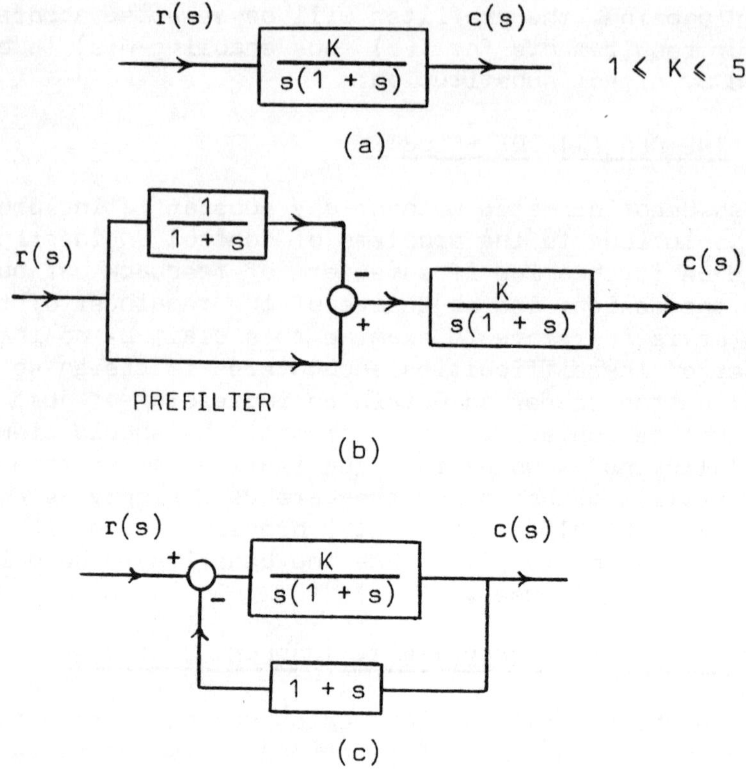

Fig (3): Compensation of a variable plant via open-loop prefiltering and feedback to achieve a given system function.

Although the systems of figure (3) have nominally the same transfer function, it is only when K is allowed to vary that the true difference between the two systems is seen. For instance, when r(t) is a unit step demand, the response for the prefiltered system is:

$$c(t) = K(1 - te^{-t} - e^{-t})$$

while that for the closed-loop system is:

$$c(t) = 1 - \frac{K}{a(b-a)}e^{-at} - \frac{K}{b(a-b)}e^{-bt}$$

It is immediately obvious that the final steady-state responses are widely different. Whereas that of the closed-loop system approaches the demanded value of unity according to the exponential modes a and b, the response of the open-loop system approaches the current value of K. The complete envelopes of the transient responses for both realizations of T(s) are shown in figure (4).

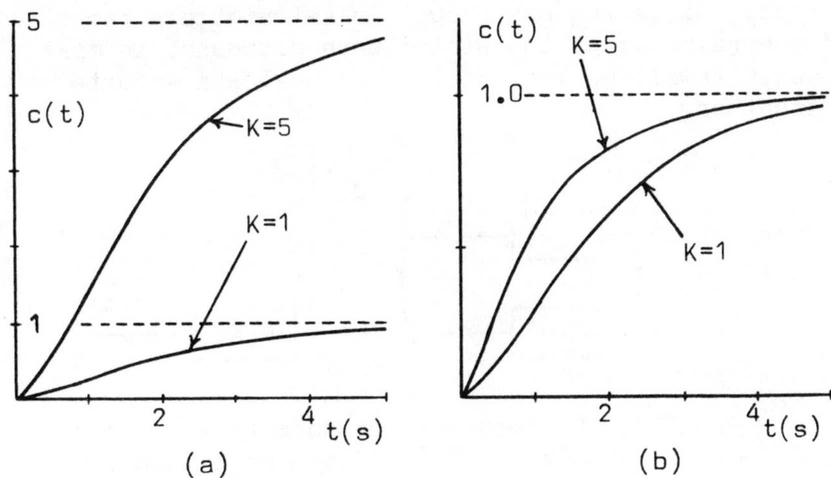

Fig (4): The unit step responses of an open-loop (a) and closed-loop system (b) for their extreme values of plant gain K.

We see from figure (4a) that the open-loop structure is totally ineffective with regard to plant sensitivity and that the proportional changes of the plant are exactly mirrored by those of the controlled system. On the other hand, feedback has greatly reduced the effect of the varying K, so much so that any response within the range of K would be acceptable.

1.5. DISTURBANCE REJECTION

It is often the case that the plant is influenced by extraneous events which cause the controlled variable or output of the system, c(t), to differ from that which was required. In some situations the disturbances are those which are to be expected during the normal operation of the plant, as when an increased load demand is made upon say a power-supply system causing the output to fall. If the disturbance can be measured then it may be possible to use the measurement as an extra input to the system to counteract the effect of the disturbance. This technique is often referred to as 'disturbance feed-foward' and is illustrated in figure (5) in which the disturbance is filtered prior to its application to the forward-path shaping network. Suitable filtering can result in satisfactory reduction of the response to the disturbance although, in most cases, it will be impossible to cancel its effects altogether.

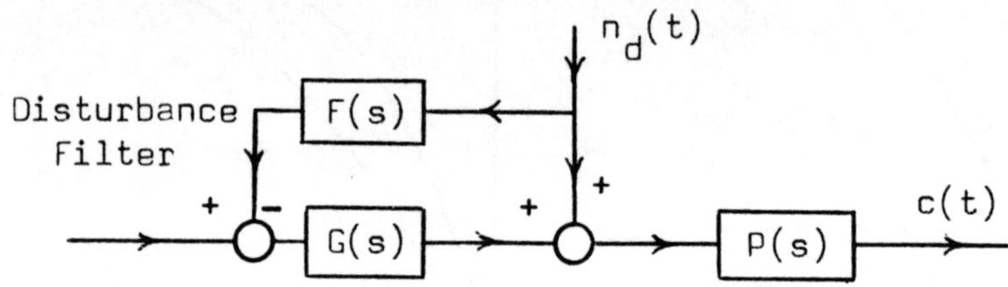

Fig (5): Disturbance feedforward - a means of attenuation of the effects of external measurable disturbances.

In other situations however, the actual disturbance may be unknown and/or unmeasurable as with the effect of sea-state on the yawing motion of a ship under auto-pilot control. Use has then to be made of the only source of system knowledge, namely the outputs, to effect a reduction of the disturbance reponse - ie, we are obliged to invoke a feedback structure.

Let us compare the performance of the open and closed-loop systems of figure (6) with respect to their ability to reject, or at least to attenuate, a noisy external disturbance. The block diagrams are those of figure (1) and (2b), rearranged to represent n_d as the system input with $r(t)$ and $n_i(t)$ both equal to zero. If the transfer functions of the open and closed-loop systems are denoted by $T_o(s)$ and $T_c(s)$ respectively, then:

$$T_o(s) = P(s) \quad : \quad T_c(s) = \frac{P(s)}{1 + L(s)}$$

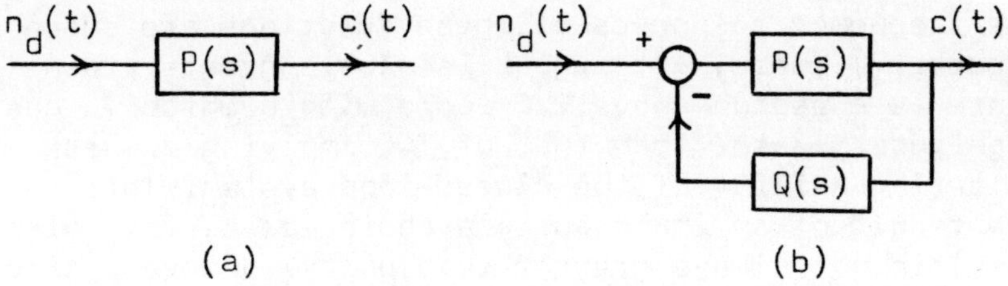

Fig (6): Open and closed-loop configurations drawn to illustrate n_d as a system input. Note that $Q(s) = G(s)H(s)$.

The two functions T_o and T_c can be looked upon from the frequency response point of view as two low-pass filters which process the disturbance. Their effectiveness in rejection will depend upon how rapidly their frequency responses fall towards zero with increasing frequency. However, since T_c is related to T_o by the term $(1+L)$, it will be obvious that T_c will be more effective in this role provided $(1+L)$ is always greater than unity. As an example, suppose:

$$T_o(s) = \frac{1}{s(1+s)} \quad : \quad L(s) = \frac{10(1+0.153s)}{s(1+s)}$$

$$\ldots \text{ for which, } \quad T_c(s) = \frac{1}{s^2 + 2.53s + 10}$$

Fig (7): Comparison of the filtering ability of the open and closed-loop functions T_o and T_c.

The frequency responses of these functions are shown in figure (7) where, although L is always greater than T_o, there is a region above 2.6 rad/s within which T_c has a magnitude greater than that of T_o, implying a worse filtering ability of the closed-loop system. This is the result of an inappropriate choice of L, for which (1+L) is not always greater than unity. However, since the attenuation of the disturbance in this region is at least 17.5db, the difference is of little concern. The superiority of the closed-loop system in reducing the effect of disturbances is high-lighted in figure (8) which shows the responses of both systems to a random disturbance. The inferior filtering action of T_o at the lower frequencies is apparent in the large variations about zero of the output response. It only remains now to decide the overall transfer function T(s) relating the actual system input to the closed-loop output, which is given by:

$$T(s) = \frac{G(s)P(s)}{1 + L(s)} = G(s)T_c(s) \qquad 1.4$$

The advantage of the two degree-of-freedom is noticed in that opportunity is given to select, independently, both the loop function to achieve the required disturbance attenuation and the forward-path function G(s) to shape the form of T(s) which decides the dynamic performance.

It is noticeable from figure (7) that even with the use of feedback T_c will transmit all low-frequency noise, albeit with a constant level of attenuation. This is particularly important when the disturbance takes the form of a constant load demand which would result in a constant offset of the output. A loop function is required which constrains the frequency response of T_c to fall to zero with decreasing frequency so that there will be no transmission of zero-frequency disturbance. This is arranged by incorporating an integrator into G(s) together with suitable stabilizing compensation. This has the action of inserting a free s-term into the numerator of T_c which will then have the required low-frequency response. Use of this technique of rejection is known appropriately as integral control.

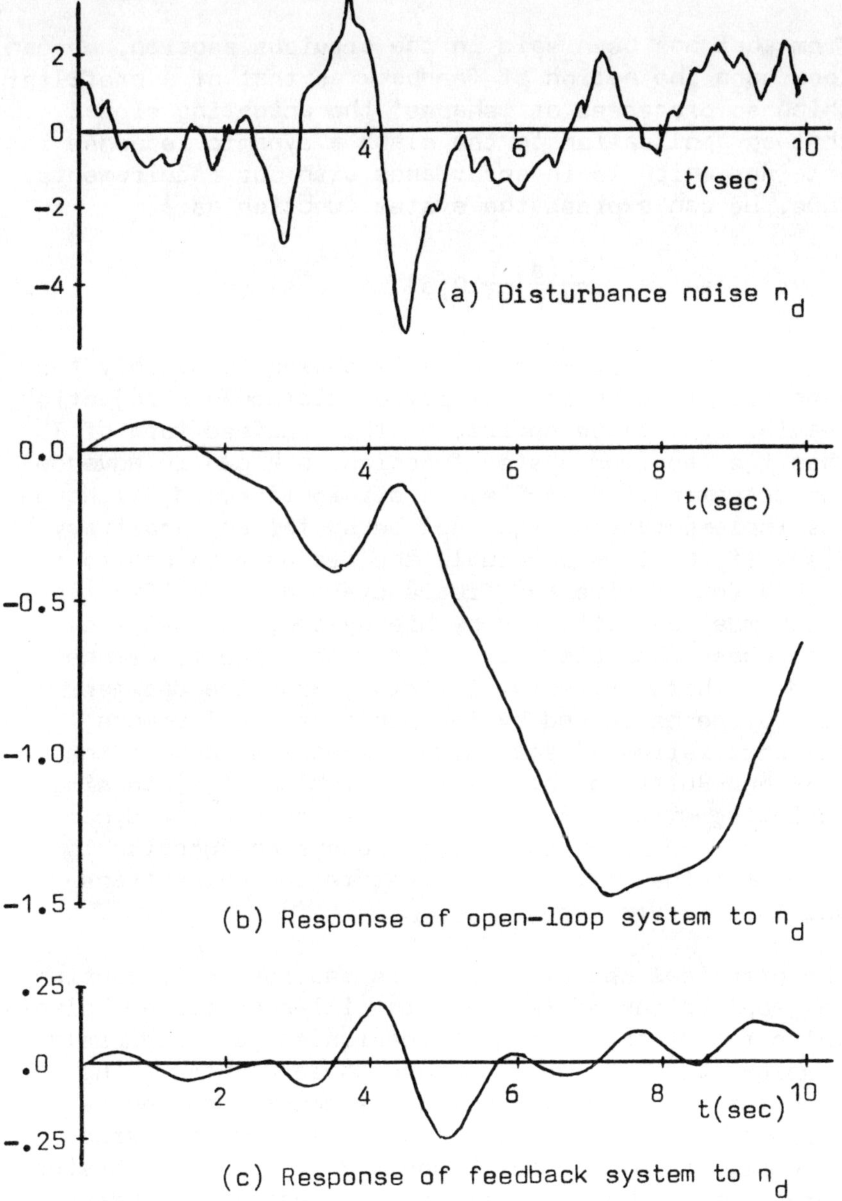

Fig (8): A comparison of the responses of an integrating open-loop plant and corresponding closed-loop configuration to an external noise disturbance n_d.

1.6. ACHIEVEMENT OF SATISFACTORY DYNAMIC RESPONSE

From what has been said in the previous section, we can look upon the action of feedback as that of a prefilter which so processes or 'shapes' the actuating signal that on application to the plant a dynamic response is obtained which is in accordance with our requirements. Thus, we can express the system function as:

$$T(s) = \frac{G(s)}{1 + L(s)} P(s) = F(s)P(s) \qquad 1.5$$

.. where L(s) is chosen to satisfy the specifications on sensitivity or disturbance rejection, leaving G(s) to be decided by the required form of T. When the required system function is known in advance the determination of G(s) is straightforward, although its implementation might not be so for any arbitrary T(s). If, as is more usual, the designer is confronted with a set of time and frequency-domain specifications which must be satisfied by his system, he must seek a T(s) whose characteristics lie within the tolerance band of those criteria. In this search the designer will often be guided by the published performance characteristics of standard-form models such as the ITAE and Butterworth. His is a problem of synthesis, beginning with a dominant model to ensure the dynamic response and later expanding the system function to cater for the desired steady-state and noise transmission performance.

The principal objective of this section is to consider the application of feedback to 'ill-conditioned' plants which render the subsequent design stages needlessly complex. Within this group are those plants having transfer functions which include non-minimum phase terms (NMP), that is, with some poles and/or zeros lying in the right-half plane (RHP), and those having very lightly-damped secondary resonances which give rise either to highly oscillatory modes of vibration or whose poles readily migrate into the RHP. Such plants are certainly not uncommon as instanced by the large antenna structures of tracking systems, by resilient hydraulic drives and by plants which are either unconditionally unstable (some ships and high-speed air craft) or have a propensity towards instability such as boilers. By far the easiest approach to system design is to begin with the plant and to modify it by minor-loop feedback to remove any undesirable characteristics whilst ensuring to maintain its steady-state potential. In particular, the designer must resist the temptation to demand more of the compensated plant than the plant of itself can give.

Let us begin the examination of the effectiveness of feedback in these situations by considering plants of varying order but with each having one pair of poles which lie close to the imaginary axis of the s-plane as shown in figure (9). The problems associated with such plants are ones of instability even with low loop gains and the presence of slowly decaying high-frequency oscillations within the system (2). Even though this undesirable mode may be barely discernible in the output waveform, its presence within the hardware may be of such a magnitude as to give rise to concern over the day-to-day integrity, reliability and maintenance of the system. If it should appear for instance in the torque drive of a servomechanism then 'chattering' of the gear train can occur which would lead eventually to excessive backlash and unnecessary maintenance.

The tendency for such systems to become unstable upon closing the loop is well illustrated in figure (9) which depicts the root-loci of several plants having n poles and m zeros which lie in close proximity of the origin in comparison with the resonant pair. Such an exaggeration is employed solely to high-light the salient features of the root-loci and frequency responses associated with resonances and any conclusions drawn from the figure should not be used for plants whose resonances lie within, say, ten band-widths of the system. One can generalise to a certain extent however to predict that compensation via the forward path function G(s) will be less effective than that using a well chosen feedback function. The explanation lies in the fact that any lead term which is introduced into G(s) must, for reasons of physical realizability, be accompanied by a lag term which will

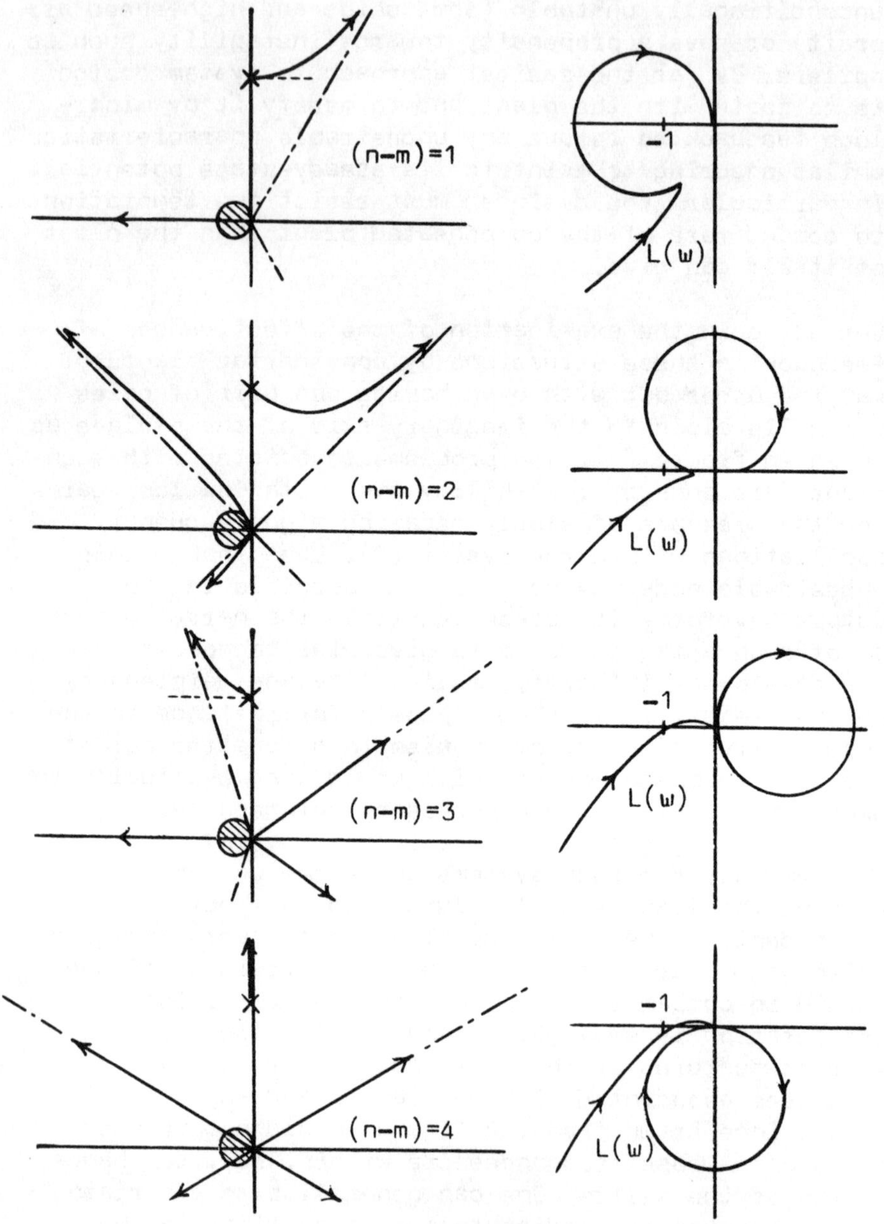

Fig (9): The root-loci and frequency responses for plants having one highly resonant pole pair. All other poles and zeros are assumed relatively close to the origin (within shaded area) with a pole-zero excess of (n−m).

tend to cancel the phase lead in the region of the resonance. On the other hand, if the feedback consists of the output and its derivatives only, they will give rise to pure zeros in the loop function without any associated poles as will be evident from the system of figure (10) which is compensated using acceleration and rate feedback.

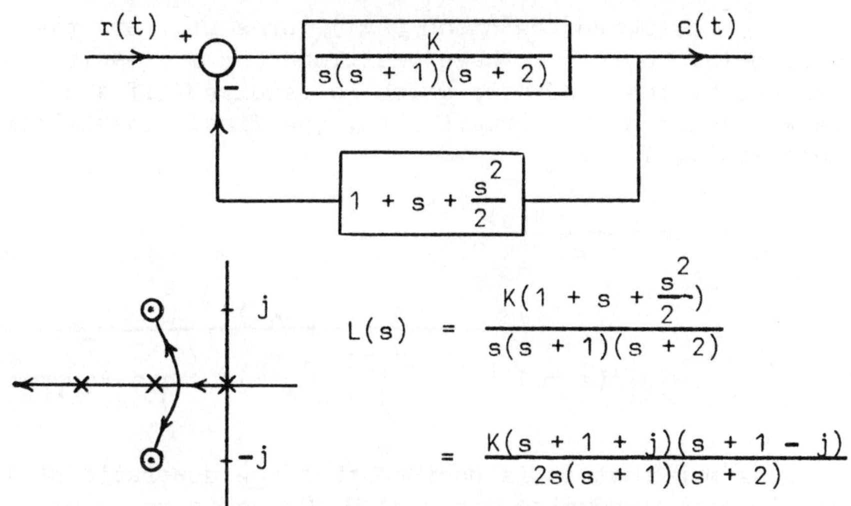

Fig (10): Zeros of the feedback function H(s) appear as zeros of the loop function. NB, they are not zeros of the system function T(s).

It should be noted that the zeros of the feedback term H(s), although appearing as zeros of the loop function, are not present as such in the closed-loop function where their effect is to modify the coefficients of the denominator to create the required poles. Thus, whilst shown on the root-locus diagram for L(s), the feedback zeros do not form part of the pole-zero array of T(s) and can, if required, be located coincident with an open-loop pole in order to pin it in that position as a closed-loop pole. Consider as an example a plant given by:

$$P(s) = \frac{1}{s(s+1)\left(1 + \frac{2 \times 0.1}{4} s + \frac{s^2}{16}\right)}$$

It is required to retain the integrating property of the plant and the pole at s = −1, whilst invoking a feedback configuration to modify the lightly-damped resonance. This can be done by emplying a third-order feedback function of the form:

$$H(s) = ks(s + 1)(s + a)$$

.. where the s and (s+1) terms will pin the closed-loop poles at these positions and s = −a will be decided by the quadratic which is required. If a gain K is employed in the forward path, the final closed-loop function will be:

$$T(s) = \frac{KP(s)}{1 + KH(s)P(s)}$$

$$= \frac{K/(1 + Kka)}{s(s + 1)\left\{1 + \left(\frac{0.005 + Kk}{1 + Kka}\right)s + \frac{s^2}{16(1 + Kka)}\right\}}$$

Suppose now that it is decided that the quadratic of T should have a damping ratio of 0.707 and a resonant frequency of 5 rad/s, then by direct comparison we find that:

$$K = 1.563 \quad ; \quad k = 0.28 \quad ; \quad a = 1.29$$

The final arrangement of plant and minor loop is shown in figure (11) together with the root-loci and the transient and frequency responses. It is very clear that the feedback has proved most effective in eliminating the sharp resonance, thus improving the relative stability of the plant whilst simultaneously removing all traces of the secondary mode from the rate response.

The success of this design lies in the use of feedback having an order only one less than that of the plant, which forces all of the poles except for one to migrate towards the zeros. The odd pole moves with increasing loop gain along the negative real axis towards infinity and takes a diminishing role in the system dynamics.

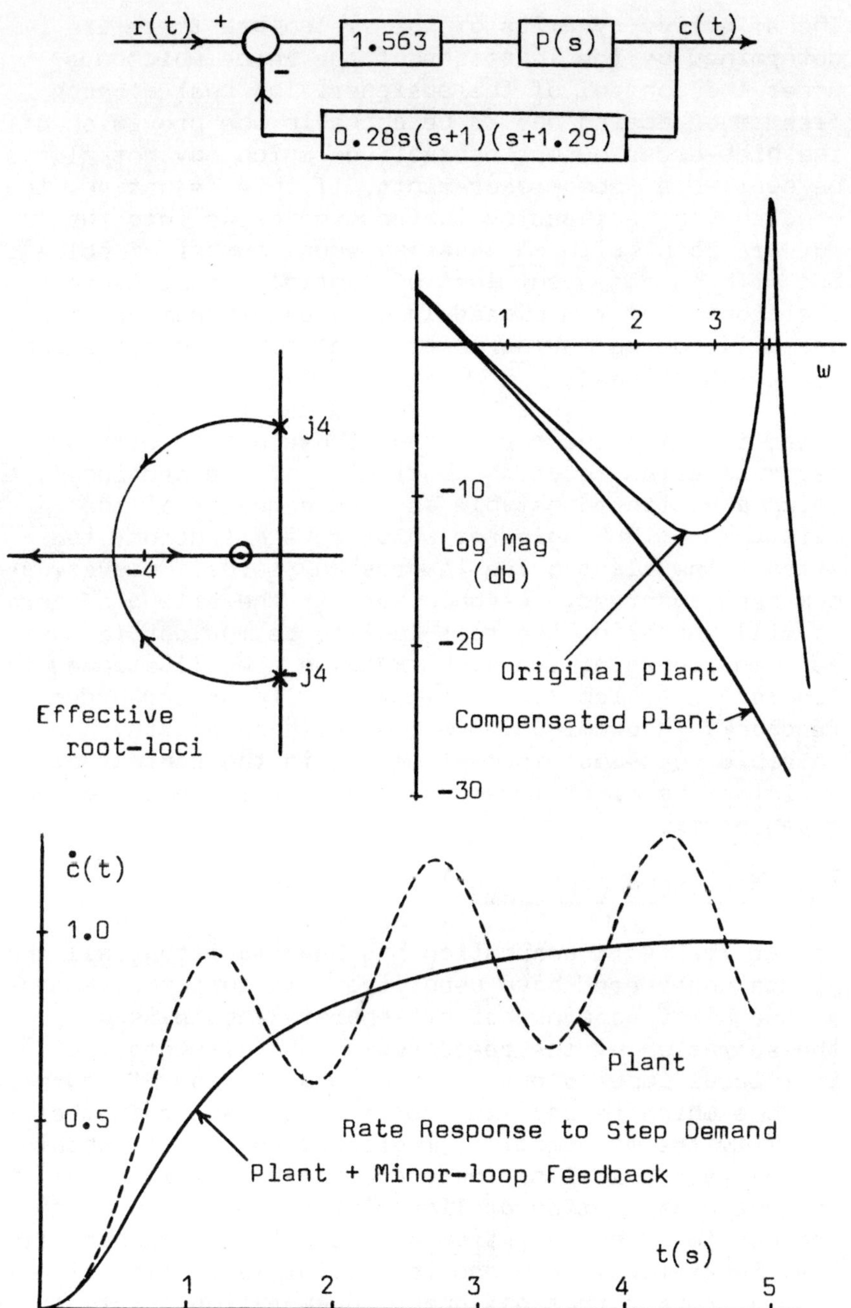

Fig (11): Improvement of plant behaviour by use of minor-loop feedback.

The effective dynamics of the system are therefore determined by the locations of the zeros which are under the control of the designer. The cost of such freedom of action has to be borne in the provision of the high-order output derivatives which may not always be available from measurements. If this is not so, the problem can be eased by introducing zeros into the forward path (with at least an equal number of poles), but with an attendant loss of control authority in the positioning of the closed-loop poles since more than one will now be moving towards infinity along the root-loci asymptotes.

There is still the question of those plants with NMP terms referred to at the beginning of the section, which are either unstable as a consequence of the existence of RHP poles or which have a tendency to attract the plant poles towards RHP zeros. However, as has been observed, feedback permits the siting of zeros at will, enabling the root-loci to be moulded to meet our requirements. Two such examples are illustrated in figure (12) which are compensated with second-order feedback. An examination of the pole-zero array and the possible root-loci is most useful in the context of deciding the minimal compensation requirements for any given plant.

1.7. NONLINEAR BEHAVIOUR

So far the tacit assumption has been made that all the plants considered have been linear in that the response of the plant to a number of separate inputs is equal to the summation of the responses of the plant to the individual input signals. This is a fiction of course, but one which is perpetuated at some time or another to simplify the mathematical manipulations and to obtain an analytic solution. There are many situations in which the assumption of linearity is justified on the grounds that the deviation of the actual responses from the linear-model response is negligible for low signal levels or perturbations about an operating point. It is inescapable however that all plants are inherently non-linear. Whether the nonlinearity will have significant effects on the performance of the system may well

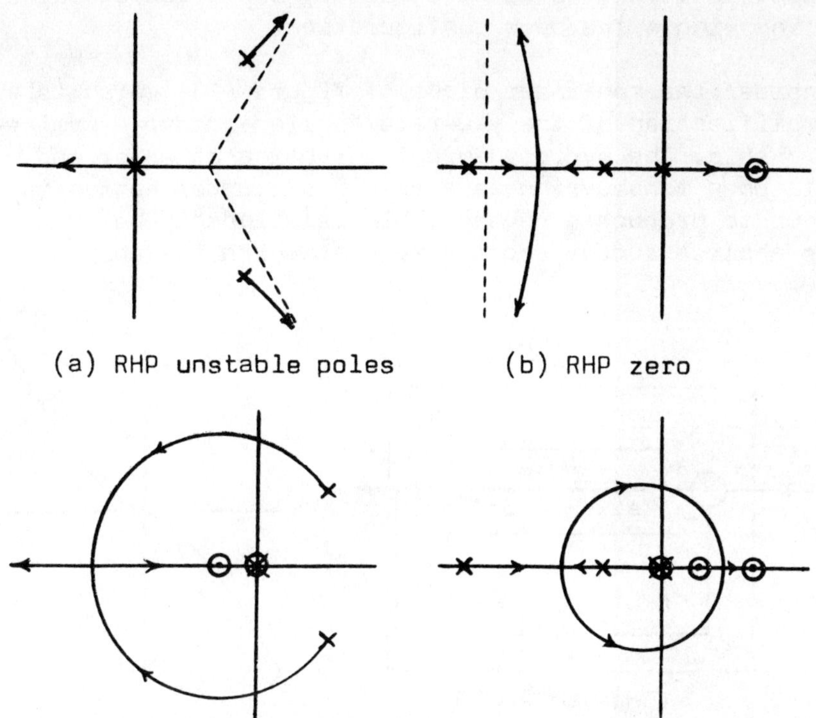

Fig (12): The stabilization of a plant (a) which is inherently unstable and a plant (b) which would be unstable on closing the loop, by the introduction of two feedback zeros. In each case, one zero is placed at the origin to ensure integration property.

depend on the demands made or the mode of operation. It has been observed earlier that the distortions of the response caused by nonlinearities can be reduced by the use of feedback, that is, the effect of feedback is to linearize to some extent an otherwise nonlinear plant. Although it should be possible to achieve linearization by means of cascaded filters with inverse nonlinear characteristics, it cannot be relied upon in every case since the exact form of the nonlinearity may be unknown or variable (this will be expanded upon later). One can think of a nonlinear element as being, conceptually, a linear element with parameters which exhibit sporadic variation. Viewed in this light, the problem of non-linear compensation appears as one of sensitivity reduction which, as we have already seen, can be solved by invoking a feedback configuration.

Consider the nonlinear plant of figure (13) which is a simplification of the yaw-rate/applied-rudder dynamics of a ship. The hydrodynamic interaction of water and hull when manoeuvring as a result of rudder action is known to produce a severe cubic relationship between the applied rudder and the rate at which the ship will yaw.

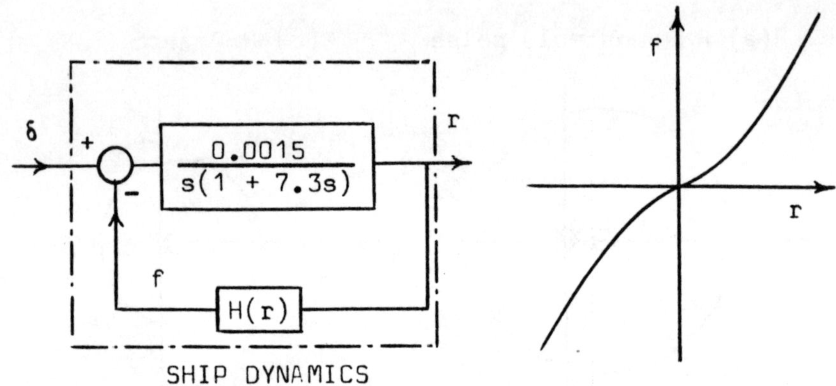

Fig (13): A schematic representation of the yaw-rate (r) to rudder angle (δ) dynamics of a ship hull with inherent 'lumped' hydrodynamic feedback.

Let it be assumed that the maximum rudder angle which can be applied is 35° with which it is required that the ship should yaw in the steady-state at 1°/s. If the nonlinear inherent feedback is defined by:

$$H(r) = 5r + 30r^3$$

.. and if feedback is applied as shown in figure (14), then we have:

$$7.3r + 1.724\dot{r} + 0.1124r = 0.0045\delta - 0.045r^3$$

Without further analysis it can be seen that the steady state relationship between r and δ is given by:

$$\delta \triangleq H'(r) = 25r + 10r^3$$

The feedback has therefore effectively reduced the degree of nonlinearity as far as the modified plant is concerned.

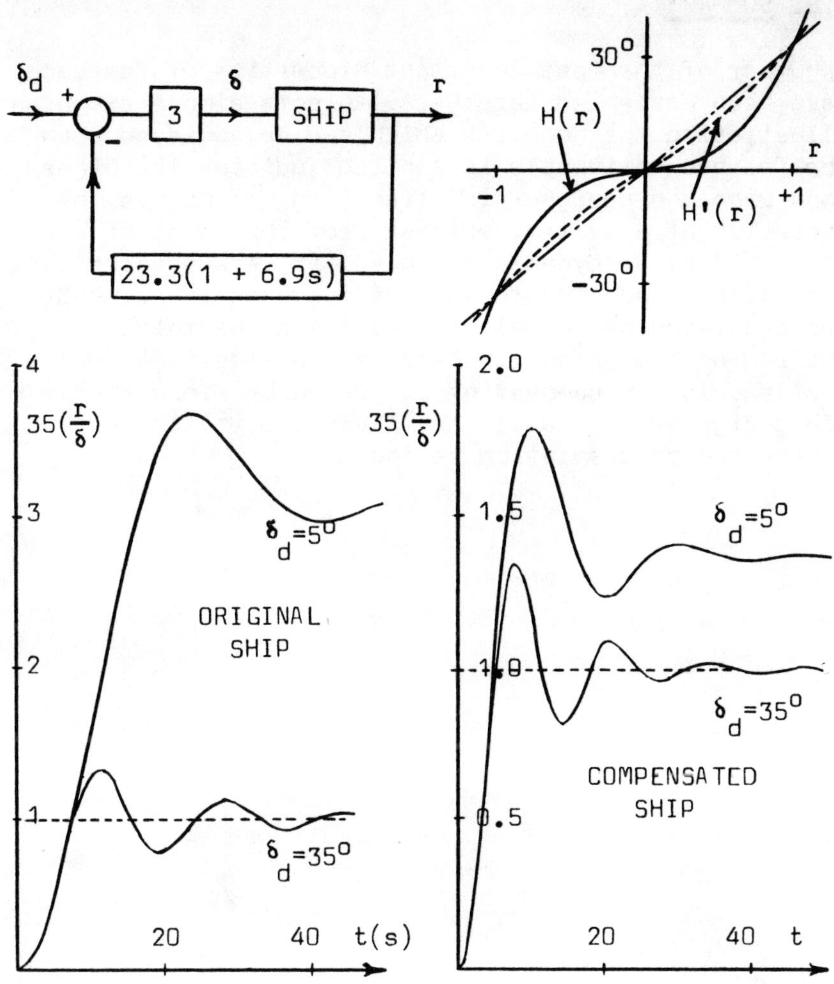

Fig (14): Block diagram of the feedback compensated ship with comparisons of the original and achieved nonlinear characteristics and the transient response to step rudder demands.

Because of the inherent nonlinearity, the modes and the magnitudes of the ship motion are seen to vary widely with the magnitude of the rudder-angle demand, whereas with feedback applied some linearization of the rudder-yaw rate characteristic is observed together with a reduction of the spread of responses.

1.8. SUMMARY

A number of the most important properties of feedback have been presented together with some simple examples illustrating the benefits which can be achieved when the feedback principle is applied judiciously. Other than when the designer is attempting to improve the operation of a system, whether from the point of view of stability or dynamic response, it is suggested that the basic reason for the use of feedback is to handle our ignorance of the plant itself and the total environment in which it operates. Suffice it to say that should the compass of ignorance be diminished, the design problem will inevitably be simplified and a more economic solution be found.

CHAPTER 2
Considerations of Dynamic Accuracy

2.1. INTRODUCTION

It will have been noticed that the previous chapter was concerned primarily with the ability of feedback to handle the vagaries of plant behaviour and that little attention was given to the overall dynamic response of the system. This order of priority is indeed correct since the plant will almost certainly have been designed without reference to the control engineer and before he is in a position to give consideration to the dynamic aspects of the system any ill-conditioning of the plant must be eliminated. It is seen therefore that there is a division of responsibility in the conception of the system with minor-loop feedback preparing the ground as it were for the following deliberations of the designer in the role of dynamic analyst. Thus, now considering the compensated plant as a stable and acceptable entity, he will set about the design of the outer or 'position' control loop.

Although not the fundamental subject of concern of this monograph, closed-loop dynamic performance is an aspect of the system which is most pertinent to the user and one therefore worthy of some attention.

2.2. PERFORMANCE SPECIFICATIONS AND TASK DEFINITION

The performance required of a system must be obtained from a thorough examination of the real-life tasks which it is expected to carry out and, from an analysis of the requirements, to provide an interpretation in a form suitable for the design approach adopted. Care should be exercised in defining the problem since the performance specification must state in concise and unambiguous terms the characteristics required if the operation of the system is to be adequate. Inability to provide a set of specifications which are compatible with one another, or to succumb to the temptation of over-specification can only result in an uneconomic design.

It is assumed that specifications having particular relevance to the selection of system hardware have been satisfied bearing in mind the economic and availability aspects. Included within these will be the maximum values of speeds, accelerations, slewing rates, peak powers and duty cycles and other items such as smoothness of motion at low speed, positioning accuracy, resonances and environmental conditions.

The aim is to identify those characteristics which are of relevance to only the short term dynamic response and which will therefore combine to define a dominant description or model of the system. It has been found by experience that the specification of system performance is most usefully accomplished by defining response criteria in the frequency and time domains.

2.2.1. TIME-DOMAIN SPECIFICATIONS

The conventional method of indicating system response is by means of the response of the system to a step-change input. Although academic in the sense that few systems will ever be expected to respond to such an input, it does provide insight into the general behaviour of the system. It is easy to apply and is sufficiently drastic in effect to excite all the modes of oscillation and hence give a measure of the speed of response and stability. By its very nature, no physical system is able to follow it, which makes it eminently suitable as a means of ranking competitive systems.

The rise time and settling time are often-quoted criteria of the response which are indicative of the system's ability to follow commands, representing the times required for the output to reach 95% of the final steady-state value and to remain within ± 5% of the same value respectively. The peak overshoot expresses unambiguously the degree of stability attained with 30-50% representing a generally acceptable performance for servomechanisms. These criteria are illustrated in figure (2.1).

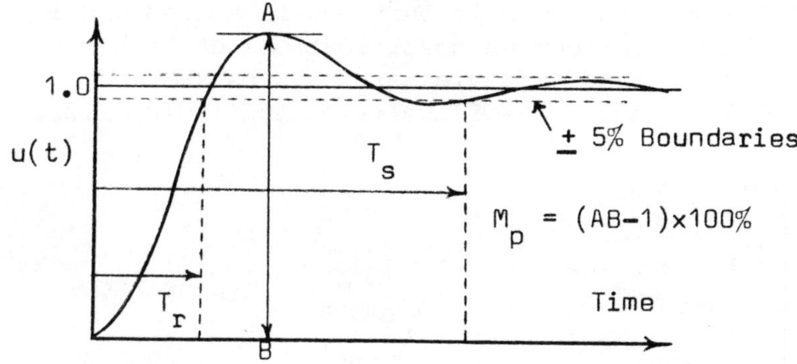

Fig (2.1): Characteristics of the unit step response - Rise Time T_r, Settling Time T_s and Peak Overshoot M_p.

The accuracy with which the output corresponds to the command input is always of concern to the designer. The bandwidth of the closed-loop system is often referred to as a measure of the ability of the system to follow the input signal, but it has been pointed out (3) that for adequate tracking the frequency spectrum of the input is only of the order of a few percent of the band width. Error coefficients are much more useful in this context. Thus, by suitable re-arrangement, the system error may be expressed as a function of the derivatives of the input:

$$\theta_e = \frac{1}{a_0}\theta_i + \frac{1}{a_1}\dot{\theta}_i + \frac{1}{a_2}\ddot{\theta}_i + \ldots. \qquad 2.1$$

where θ_e, θ_i represent system error and input and where the a_i are the relevant dynamic error coefficients.

If it is assumed that the first three terms of equation (2.1) are of greatest significance then a measure of the dynamic accuracy of the system is given by maximum values of the error in response to a step input, ramp input and parabolic input.

2.2.2. FREQUENCY-DOMAIN SPECIFICATIONS

Another way of characterizing the behaviour of a system is by the manner in which it responds to a sinusoidal command signal. The common specifications of this domain are bandwidth, phase and gain margins, resonant peak and resonant frequency as illustrated in figure (2.2).

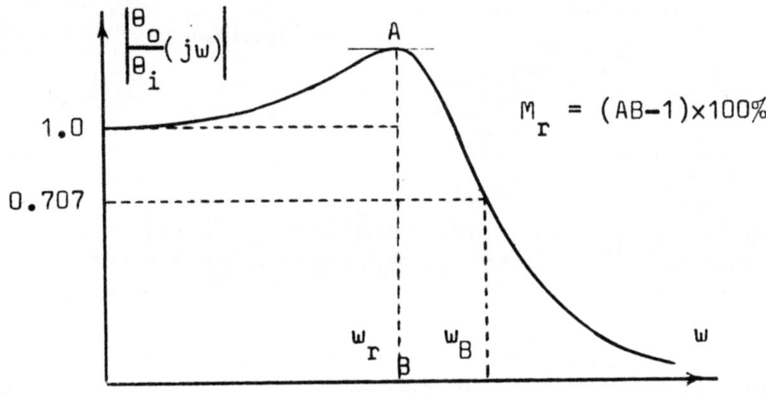

Fig (2.2): Characteristics of the frequency response - Bandwidth w_B, Resonant Frequency w_r and Resonant Peak M_r.

Ideally, a system should transmit with negligible phase and amplitude distortion all frequency components contained in the command signal for the faithful reproduction of the signal. This would require a system with infinite bandwidth. On the other hand, the ability of the system to reject noise contamination and the requirement for an economic design both demand a low bandwidth. The latter point is made evident when it is realized that the power capability demanded of the prime mover increases with approximately the cube of the frequency of the input.

If the input signal is considered in terms of its power spectrum then an improvement can be made on bandwidth

as a measure of signal transmission by estimating the output power of the system for a white-noise input. Such a noise-transmission factor has been proposed (4) which is equally useful as bandwidth for low values of M_p, but which provides a far truer evaluation of transmission for less damped systems.
The resonant peak M_r is closely correlated with M_p and is therefore a measure, along with phase and gain margin of the relative stability of the closed-loop system. Too high a value for M_r may give rise to vibrations of such proportions within the hardware (eg, the gearbox) as may lead to much unnecessary maintenance.

2.2.3. PLANT-INPUT EXCITATION

The power elements of all systems are limited in the maximum effort of which they are capable and will therefore saturate if too high a plant-input signal is applied, leading to a deterioration of performance.
This is of particular importance if noise contamination enters via the input or transducers and is especially so if its spectrum is at all wide. The noise which is transmitted to the output is, in a sense, only the tip of the iceberg since as the plant is a low-pass element the noise at the plant input will be very much higher. This then is yet another reason for opting for the least acceptable bandwidth. Indeed, it has been shown (5) that for a system of given pole-zero geometry the tendency to saturation is increased by the ratio x^{n-m} if the bandwidth is increased by x times, where (n-m) is the pole-zero excess.
This topic of noise present within the feedback loops is one of major importance to which further consideration will be given later in the text.

2.3. THE DOMINANT SYSTEM FUNCTION

The implication of the title of this section is that there is normally a least-order representation of the closed-loop system which is dominant in determining the dynamic response of the system to given stimuli. That is, the least-order equivalent (or model) is considered to be satisfactory for the adequate prediction of dynamic performance under conditions of

interest to the designer. It is on the basis of such a model that he will select parametric values such that the best trade-off will be achieved between the possibly conflicting specifications as defined by the criteria of section (2.2).
Control theory is resplendent with suggestions as to just how the selection can be made. Such techniques invariably set out to achieve some very specific objective and adjust the variable parameters to this end, as for example the specification of the modes of vibration of the system. Cut-and-try methods are also very popular in the more elementary texts. This is not to denounce such approaches to system design but rather to suggest that if specifications can be set in terms of criteria already discussed, and such a practice is widespread, then these techniques will not be appropriate.
What is suggested is the use of standard forms (6,7) for which all the information required by the designer is either already available or is readily obtained by computation. Even so, the design process can be tedious when a large number of individual specifications is involved or when, as can often happen, there is some degree of incompatibility between them. In these cases it is far easier to select the parameters of the least-order system by use of an optimization routine capable of handling nonlinear functions (8,9).

The suggested design procedure is therefore considered as a two-stage process. Firstly, the plant is modelled to include all uncertainty and disturbance and is then compensated via minor-loop feedback according to methods to be described in later chapters. Secondly, a dominant loop model is selected to suit the nature of the final system with parametric values adjusted to enable the satisfaction of the dynamic specifications. Thus, on assumption of the nominal compensated plant function, the forward and feedback-path compensation for the outer loop is obtained by straightforward comparison of terms. It may indeed be necessary to repeat these stages since the compensation functions so computed may be unnecessarily complex and be in need of reduction themselves if the compensation is to be parsimonious and realizable. Additionally, the need for adequate steady-state accuracy and high-frequency noise attenuation may call for the inclusion of dipoles and far-off filtering poles into the loop function (9).

2.4. SUMMARY

The discussion of this chapter has highlighted the primary objective of this monograph to be the presentation of design methods particularly, although not exclusively directed towards the elimination of those undesirable effects brought about by the unalterable features of real hardware plants such as parametric variation, unsatisfactory dynamics and external disturbance. The subject of closed-loop dynamic performance is essentially the province of linear dynamic analysis albeit given a logical directive by the study of reduced-order standard forms of transfer functions.

It has been suggested that these two aspects of design are sufficient to achieve a competent closed-loop system when specifications are set in terms of familiar and classical criteria such as bandwidth and peak overshoot.

System modelling has outcropped regularly throughout the discussion with special reference to reduced or low-order modelling. The reduction of complexity which can follow in its wake is possible in view of the well-known fact that many high-order systems include modes which contribute insignificantly to the overall system response. This most important topic will be reviewed in the next chapter.

CHAPTER 3
Low Order Modelling

3.1. INTRODUCTION

The modelling of physical processes is a most important subject within the field of control engineering, representing as it does the corner-stone upon which future synthesis of a viable engineering system rests. When one considers the vast array of systems to which control theory is applied, from the manifestly engineering process through biological systems to economic and managerial situations, it is also apparent that it can be a very specialised and difficult topic. It is however an attempt to represent some physical process in a mathematical manner such that the behaviour of the process in response to certain stimuli is predictable, thus enabling the process to be brought within and handled by the methods of engineering science. Widespread interest has been shown over a number of years in the associated subject of low-order modelling, by which is meant the representation of the system within a minimal mathematical framework, with the result that many ingenious techniques have been reported in the technical literature (10, 6). It is pertinent therefore to ask the question why, given all the effort which was devoted to the development of the original model, the use of a low-order model is considered at all necessary. A variety of reasons can of course be put forward, largely dependent on the problem in hand, but some of the most important are presented in the following section.

3.2. REASONS FOR LOW-ORDER MODELLING

(a) To simplify understanding of system.

It is a matter of experience that the sets of high-order equations which are obtained for real systems are very often cumbersome for design or analytical purposes and can yield a somewhat confused view of the processes involved. It would therefore be desirable to reduce the complexity to manageable proportions if at all possible. Further, it is often observed that the response of a high-order system may correspond closely to that of some much simpler system. It is therefore advantageous for design and as an aid to understanding, to employ a lower order model which describes adequately the dynamics of the process such that the salient features are clearly reflected in the model.

(b) To reduce computational requirements.

The use of low-order models must imply the attendant benefits of reduced time involvement and effort on behalf of the programmer as well as the obvious reduction in disk storage and computational time. One particular example is the saving which is possible with processes having a wide range of time constants where the computational interval must be selected according to the smallest of the set. These are just the terms which tend to be eliminated or absorbed in the model-reduction exercise, allowing simulations to be pursued in a shorter span of time. A related problem is that of sensitivity analysis where the equations involved are at least twice the order of the system and investigations into the selection of optimum parameter values may involve thousands of simulation runs. Not least, reduction of order widens the scope of problems which can be handled on a given micro-processor.

(c) To generalise system characteristics.

It will be shown later that the dynamic characteristics of any system are, to a large extent, determined by the nature of the loop frequency response in the neighbourhood of unity gain. If in addition we note that stability constraints must be imposed in this region, then it will be apparent that the scope for variation between systems is quite limited. The implication follows that a number of nominally different high-order systems may, on the basis of <u>dynamic response</u>, be considered as variants of one particular low-order model. Alternatively, a given low-order model can encompass the essential characteristics of a range of physical processes which differ primarily only in time scale. Thus, if the characteristics of the model are known and if a particular system can be associated with the model, then a change of time or frequency scale only provides an estimation of the behaviour of the original system.

(d) <u>To make best use of scanty experimental data</u>.
An alternative to the modelling of a plant is to identify its structure and estimate the parametric values experimentally via techniques such as harmonic analysis, correlation or time-series analysis. It may be the case however that for some reason the collected data may be scanty. In such a situation it is better to estimate the few factors of a low-order model with confidence rather than endeavouring to estimate many parameters with less confidence.

(e) <u>To reduce hardware requirements</u>.
The proliferation of micro-processors has encouraged the application of control strategies which once would have been considered unrealizable or too expensive. Even so, the storage capacities of these devices are strictly limited, such that the use of low-order models may be essential by necessity. The need to limit the dimensions of the process becomes even more critical with designs based on optimal theory with their implied estimation requirements making heavy demands upon the computational facilities.

Although the model should be capable of providing an adequate prediction of the system response of interest to the designer, it must not be forgotten that the reduction process clearly entails some loss of information which must result in some deviation between the predicted and actual responses. For instance, a third-order model of a plant which possesses both primary and secondary resonances could not be expected to predict the secondary-mode oscillation which appears so prominently in the impulse response.

3.3. MODELLING FROM THE LOOP FREQUENCY RESPONSE

The modelling technique to be described is perhaps the most appealing in its obvious simplicity and effectiveness of all the methods to be discussed. The user is immediately aware of the important system parameters, the minimum order of the model for good prediction and in addition is able to estimate the system poles on closing the loop. The initial impetus came from Axelby (11) and Biernson (12) whose work was extended by Kan Chen (13) who recognised than an approximate correlation existed between the behaviour of the magnitude of the loop frequency response within three regions of the plot and the closed-loop poles. With reference to figure (3.1) the regions are defined as:

Region I

This is the region of high loop gain which is dominant in deciding the steady-state accuracy of the closed-loop system and which determines the settling time should a positive break-point be evident in the asymptotic Bode plot (ie, an open-loop zero close to the origin of the s-plane). It is obvious that for high loop gain a closed-loop pole will migrate to within close proximity of the zero and will be located approximately at:

$$-\omega_z\left(1 \pm \frac{1}{x}\right) \qquad 3.1$$

where ω_z is the location of the zero and x is the linear gain at ω_z.

To avoid long 'tails' in the transient response due to the dipole, it is necessary that the gain x should be as high as possible.

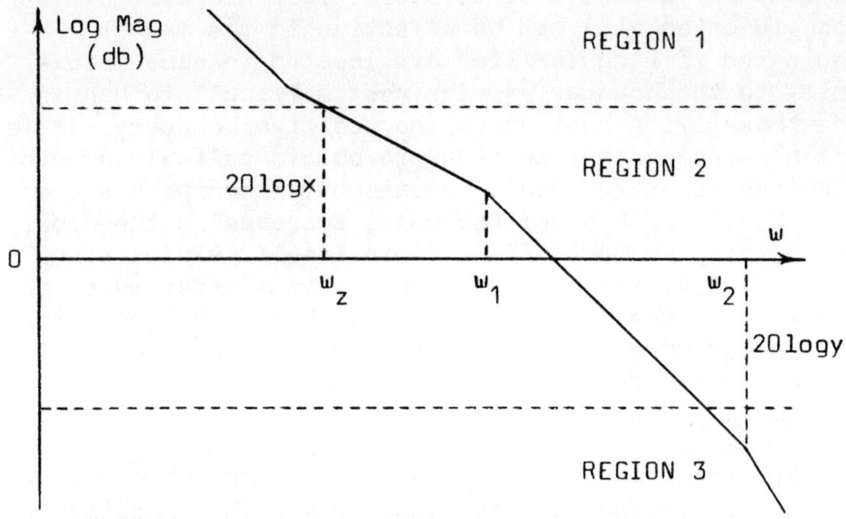

Fig (3.1): Division of the loop frequency response into three regions according to amplitude ratio and relevance to system performance.

Region 2

The region of intermediate gain which contains the dominant transfer function in that singularities in this region are dominant in determining the dynamic response and will therefore be used to constitute the low-order model.

Region 3

The region of low gain and high frequency in which the system follows very badly and which contains the 'far-off' poles and zeros whose effect is to introduce a time delay into the initial portion of the transient response and to affect noise rejection. They have little effect on dynamic errors and result in closed-loop poles located approximately at:

$$-w_2(1 \pm y) \tag{3.2}$$

where w_2 is a location of a far off pole and y is the linear gain at w_2.

It will be obvious that this arbitrary division of the loop function will not be effective in the manner indicated if singularities are located in close proximity to the boundaries. The most difficult to handle are those lying just above the positive boundary. It is often the case that in order to obtain reliable prediction the boundary should be raised to incorporate the singularity with a corresponding increase in the order of the reduced model. This is in itself helpful since it is a positive indication of the minimum order which can be used for the model. It would be known that anything less would have serious limitations. Values adopted for the boundary levels commonly lie within the range of $\pm$ 15db to $\pm$ 20db.

Far-off poles and zeros on the other hand can be dealt with in a straightforward manner (14). The objective is to replace all far-off singularities by absorbing them into the dominant transfer function of Region 2 or by creating an extra singularity which has the same phase effect at the cross-over frequency as do all those it replaces. The cross-over frequency is chosen as the reference frequency since it is most intimately related to the dynamic response.
Thus, if w_c is the cross-over frequency and all singularities are considered far-off then their phase contribution at w_c is given by:

$$\phi = -\sum_p \tan^{-1} w_c T_p + \sum_z \tan^{-1} w_c T_z$$

$$\cong -w_c\left(\sum_p T_p - \sum_z T_z\right)$$

where the summations are over all time constants T_p, T_z of poles and zeros.

The effective time constant T_e is therefore:

$$T_e = \sum_p T_p - \sum_z T_z \qquad 3.3$$

Similarly for a quadratic if the resonant frequency ω_n is much greater than ω_c then its effective contribution to the time constant is:

$$T_e = \frac{2\zeta}{\omega_n} \tag{3.4}$$

As an application of the technique consider the case of an aircraft blind-landing system (14) having the transfer function:

$$\frac{y_o}{y_e} = \frac{0.0185(1 + 10.0s)}{s^2(1 + 0.5s)(1 + 0.73s)^3} \tag{3.5}$$

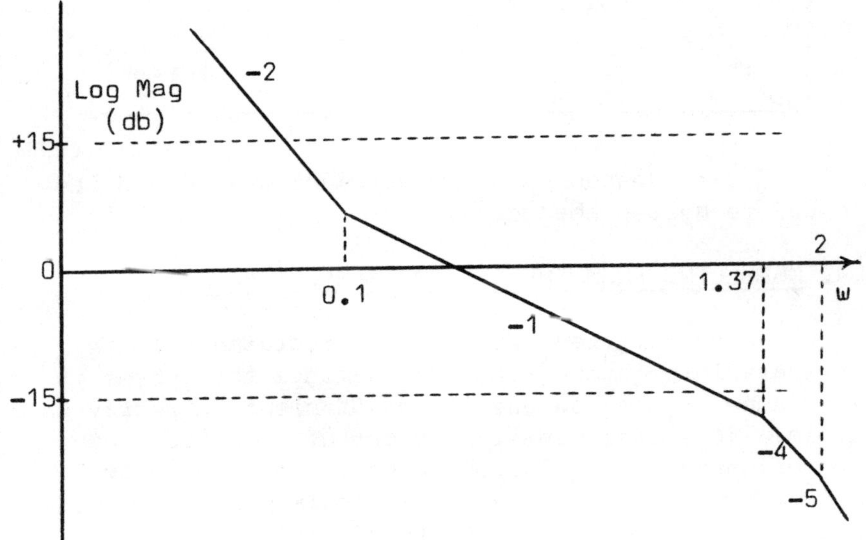

Fig (3.2): Asymptotic Bode plot of the loop function of an aircraft blind-landing system.

The effective time constant of the far-off poles is therefore given by:

$$T_e = 3 \times 0.73 + 0.5 = 2.69\text{s}$$

The sixth-order system can therefore be reduced to the

following third-order model:

$$\frac{y_o}{y_e} = \frac{0.0185(1 + 10s)}{s^2(1 + 2.69s)} \qquad 3.6$$

For comparison purposes, the step responses of the actual system and the model of equation (3.6) are shown in figure (3.3).

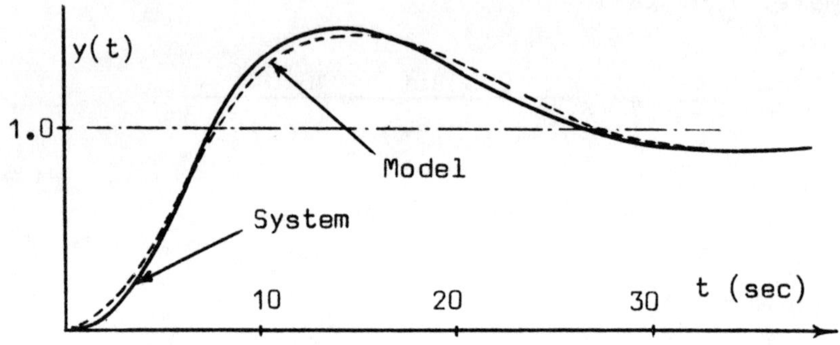

Fig (3.3): Responses of an autoland system and its model to a step change.

3.4. MODELLING FROM THE SYSTEM TRANSFER FUNCTION

Since the system transfer function contains all the necessary information required to model the system it would seem natural to use the coefficients directly to obtain a low-order model. A number of techniques have been proposed (15-19) of which only the time-delay (15) and the continued fraction (16) models will be considered as representatives of this class.

3.4.1. The Matsubara Time-Delay Model

It is well known that the contribution of the far-off poles to the transient response is negligible except insofar as the initial portion of the response is time delayed. Matsubara (15) proposed that an equivalent time-shift be introduced as a replacement for the far-poles and zeros, with the remainder of the system function forming the linear portion of the low-order model. He defined the equivalent dead-time D as follows:

$$D \triangleq \int_0^\infty \{u(t) - r(t)\}\,dt = \lim_{s \to 0} \frac{1}{s}\{1 - T(s)\} \quad 3.7$$

where $u(t)$ is a unit step function, $r(t)$ is the system step response and $T(s)$ is the system transfer function.

Thus, the equivalent dead-time is the integral of the step response error over all time which, as proposed, is equated to the overall delay introduced by the far-off singularities.
Let the system function be defined:

$$T(s) = \frac{1 + a_1 s + a_2 s^2 + \dots\dots}{1 + b_1 s + b_2 s^2 + \dots\dots} \quad 3.8$$

Substituting (3.7) into (3.8) we have:

$$D = b_1 - a_1 \quad 3.9$$

Since for the numerator and denominator polynomials of equation (3.8) the terms a_1 and b_1 represent the sum of numerator and denominator time constants, the equivalent dead time is expressed alternatively as:

$$D = \sum T_p - \sum T_z \quad 3.10$$

It is found that in general this technique provides an excellent prediction of the transient response when the order of the dominant function is chosen correctly. Unfortunately no guidelines can be given as to this choice. As examples of what can be achieved, consider again the autoland system of section (3.3).

(i) 2nd-order model:

$$T_m(s) = \frac{e^{-5.05s}(1 + 10s)}{1 + 4.95s + 22.68s^2} \quad 3.11$$

(ii) 3rd-order model:

$$T_m(s) = \frac{e^{-1.60s}(1 + 10s)}{1 + 8.4s + 39.6s^2 + 77.5s^3} \qquad 3.12$$

The step responses of the above models and the actual system are shown in figure (3.4).

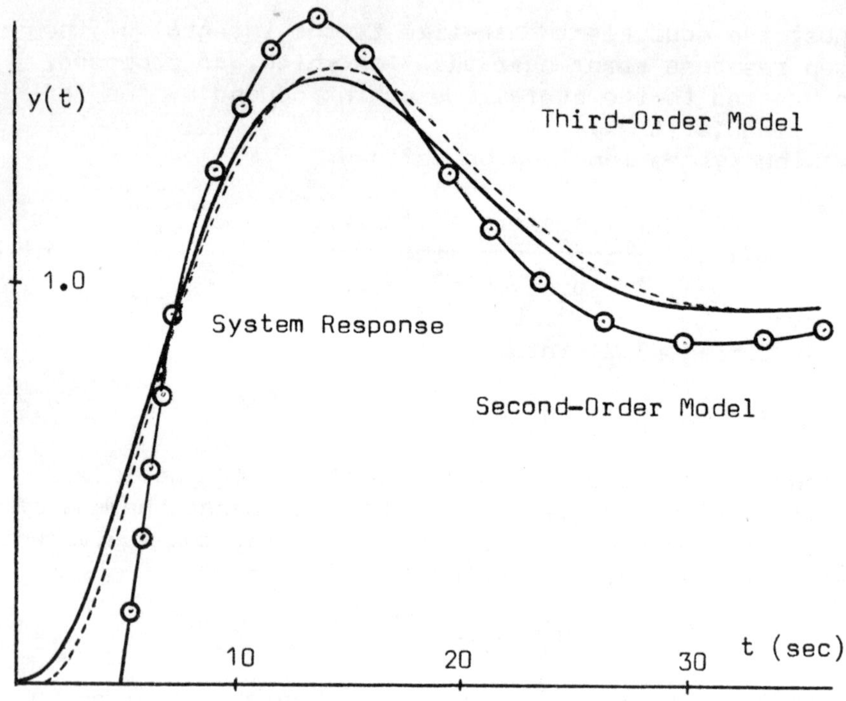

Fig (3.4): Responses of the Matsabura models of an autoland system in comparison with the actual system response.

3.4.2. The Continued-Fraction Model

Chen and Shieh (16) developed a simple method of reduction based on the expansion of the high-order function as a continued fraction which is truncated to yield the low-order model. The quotients in the expansion are in order of decreasing significance of their contribution to the system response and have the valuable property of converging rapidly. The model does not retain the dominant poles of the original system but modifies the structure such that the pole-zero excess is always unity. The technique is well suited to digital computation. Thus, let the original system have the following transfer function:

$$T(s) = \frac{a_{2,1} + a_{2,2}s + \dots + a_{2,n}s^{n-1}}{a_{1,1} + a_{1,2}s + \dots + a_{1,n+1}s^{n}} \qquad 3.13$$

By repeated division we have:

$$T(s) = \cfrac{1}{\cfrac{a_{1,1}}{a_{2,1}} + \cfrac{s}{\cfrac{a_{2,1}}{a_{3,1}} + \cfrac{s}{\cfrac{a_{3,1}}{a_{4,1}} + \cfrac{s}{\cdots\cdots}}}} \qquad 3.14$$

The additional coefficients are determined by Routh's algorithm:

$$a_{i,j} = a_{i-2,j+1} - \frac{a_{i-2,1} \cdot a_{i-1,j+1}}{a_{i-1,1}} \qquad 3.15$$

$$\text{for, } i = 3,4,\dots\dots,2n+1$$
$$j = 1,2,\dots\dots\dots,n$$

If a model of order m is required, then 2m terms of the form $a_{i,1}/a_{i+1,1}$ are computed and the model function is reconstituted. It is always found that the numerator of the model function has an order of (m-1), which indicates that even though the predicted transient response may be satisfactory, the frequency response can be very seriously in error. The model obtained is identical to that derived using the Pade approximation technique.

Consider as an example the following system:

$$T(s) = \frac{1 + .02s}{1+.522s+.307s^2+.028s^3+.00139s^4+.00003s^5}$$

On forming the terms of the continued fraction we have:

$$T(s) = \cfrac{1}{1 + \cfrac{s}{1.992 + \cfrac{s}{-0.848 + \cfrac{s}{-2.28 + \cfrac{s}{\cdots\cdots}}}}}$$

This reconstitutes to form the second-order model:

$$T(s) = \frac{1 - 0.2108s}{1 + 0.291s + 0.2596s^2}$$

Note that the model is non-minimum phase in that the zero lies within the RHP and the effect of which can be seen in figure (3.5) where the initial portion of the response has a negative excursion.

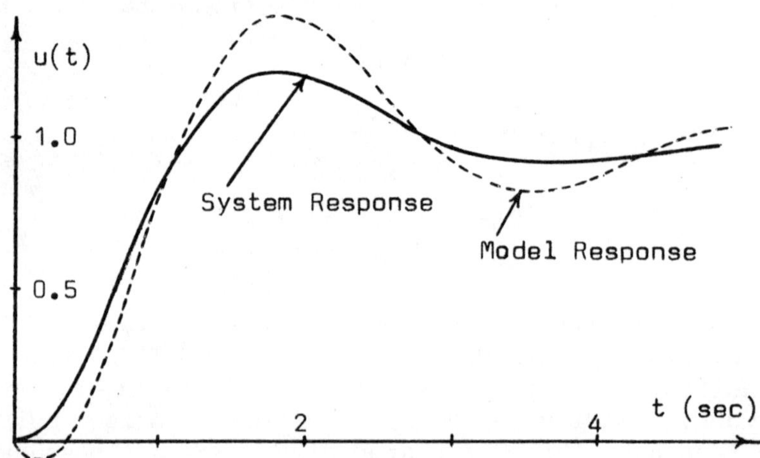

Fig (3.5): The effect of a non-minimum phase zero on the response of a continued-fraction model.

Although the response prediction of figure (3.5) is

very poor this is not normally the case with this particular technique. In general it is found that a third-order model is necessary for adequate prediction, but the second-order model was chosen above to highlight the negative excursion which is possible with the method. Indeed, it is not uncommon for it to produce models with RHP poles, that is, to give rise to unstable models from stable systems. This characteristic, together with the inferior prediction of the frequency response means that the technique should be used with care.

It can be shown that the Chen model is formulated by expanding the system function about s = 0 and by retaining only the most significant terms. An improvement was reported by Chuang (18) by expanding instead about both s = 0 and infinity, an approach which produced a better prediction but which was still bedevilled by RHP singularities. Whalley (19) solved this problem by assuming the existence of a dead-time in the system function and adjusting its value until all excess phase (as exemplified by the RHP terms) had been taken up, leaving a minimum phase model. Since the dead time $\exp(\pm sT)$ only affects the numerator, the numerator polynomial of the final model can be recalculated for the specific value of T employed.

It may be argued that providing the degree of undershoot of the transient response is small the existence of a RHP zero is unimportant. However, if the model is to be employed in the design of a closed-loop system and if the model has RHP zeros, then the closed-loop poles must migrate towards the RHP and predict instability where none actually exists.

3.5. MODELLING FROM THE FREQUENCY RESPONSE

The basic method considered in this section, and one which has been widely employed in industry, is the weighted least-squares technique introduced by Levy (20). Whereas the straightforward application of least squares to the curve-fitting problem results in quite difficult computation, Levy's change to the weighted version greatly simplified the procedure. It will be seen that the simplification is achieved only at the cost of reduced accuracy of fit and that only by the introduction of an iterative procedure is the expected accuracy restored. Define the following functions:

$F(j\omega)$ = Original function to be modelled, known in terms of its real and imaginary components at a set of frequencies ω_k.

$G(j\omega)$ = Approximating model function formed as the ratio of two polynomials $N(j\omega)$ and $D(j\omega)$.

$e(j\omega)$ = Error existing between $F(j\omega)$ and $G(j\omega)$.

We therefore have:

$$e(j\omega) = F(j\omega) - G(j\omega) = F(j\omega) - \frac{N(j\omega)}{D(j\omega)} \qquad 3.16$$

... or,

$$D(j\omega)e(j\omega) = D(j\omega)F(j\omega) - N(j\omega) \triangleq a(\omega) + jb(\omega)$$

Thus, at any specific frequency ω_k,

$$|D(j\omega_k)e(j\omega_k)|^2 = a^2(\omega_k) + b^2(\omega_k) \qquad 3.17$$

An error function E is now defined as:

$$E \triangleq \sum_{k=1}^{m} \{a^2(\omega_k) + b^2(\omega_k)\} \qquad 3.18$$

On minimisation of E with respect to all coefficients of $G(j\omega)$ a set of simultaneous equations is obtained:

$$(A).(B) = (C)$$

where B is a column matrix of unknown model coefficients and A,C are functions of the known frequency response measurements.

The simplification introduced by Levy which provides a unique solution for $G(j\omega)$ is obtained at the cost of reduced accuracy of fit at the lower frequencies since, as may be seen from equation (3.17), the actual error is weighted by $D(j\omega)$ which assumes greater proportions with increasing frequency. Such weighting is readily eliminated by the application of the following procedure (21).

Define a new error function $e'(j\omega)$ such that:

$$e'(j\omega) = \frac{e(j\omega)D(j\omega)_L}{D(j\omega)_{L-1}}$$

$$= \frac{F(j\omega)D(j\omega)_L}{D(j\omega)_{L-1}} - \frac{N(j\omega)_L}{D(j\omega)_{L-1}} \qquad 3.19$$

where $D(j\omega)_L$ is the denominator computed at the L-th iteration.

Since $D(j\omega)$ is not known initially, it is assumed for the first iteration that it is equal to unity. Now as for Levy's approach define an error function:

$$E = \sum_{k=1}^{m} \left| \frac{e(\omega_k)D(\omega_k)_L}{D(\omega_k)_{L-1}} \right|^2 \qquad 3.20$$

Minimisation is again carried out with respect to the coefficients of $G(j\omega)$ to obtain a set of equations, the solution of which gives the coefficients at the L-th iteration. It is obvious from equation (3.19) that successive iterations lead asymptotically to a true least-squares fit.

It is sometimes found that in endeavouring to obtain a low-order model from frequency response data the resulting model is found to have non-minimum phase terms as was the case with the continued-fraction model. This effect can be attributed to a single cause, namely that the low-order model is attempting to match the large phase shifts of the high-order system with the phase shift achievable by fewer terms, an objective which in certain circumstances can be met only by allowing the singularities to take up locations in the RHP. Payne (22) was able to reduce the incidence of such terms by incorporating constraints into the algorithm which were based upon a priori information of the system such as steady-state gain and error coefficients.

Consider now the application of the iterative least-squares method to the modelling of an 8th order electro hydraulic servomechanism having the following transfer function.

$$T(s) = \frac{(1+0.0541s)}{(1+0.046s+0.0014s^2)(1+0.0016s+0.000012s^2)(1+0.0039s)(1+0.002s)(1+0.0013s)(1+0.0006s)}$$

It was known a priori that the above system had a unit steady-state gain and that its dynamics approximated closely to that of a Type II system. Using this information (23), a third-order model is obtained:

$$T_m(s) = \frac{1 + 0.0532s}{1 + 0.0532s + 0.0014s^2 + 0.0000096s^3}$$

Some relevant responses of the model and the original system are shown in figure (3.6) from which it is seen that although the prediction of the step response is quite good there is a considerable divergence in the prediction of response to impulse and high-frequency sinusoids. This is a consequence in the main of the presence of a lightly-damped secondary resonance in the actual system. If a third-order model is required, there is no possibility of modelling the secondary-mode oscillations of the impulse response nor the gross phase shifts which occur at high frequencies. This is true for any modelling technique employed and should bring fully to mind the necessity for the designer to give very careful consideration to the characteristics of his system and the purposes for which he requires the model before he decides on the order of the model and the particular modelling technique to be employed.

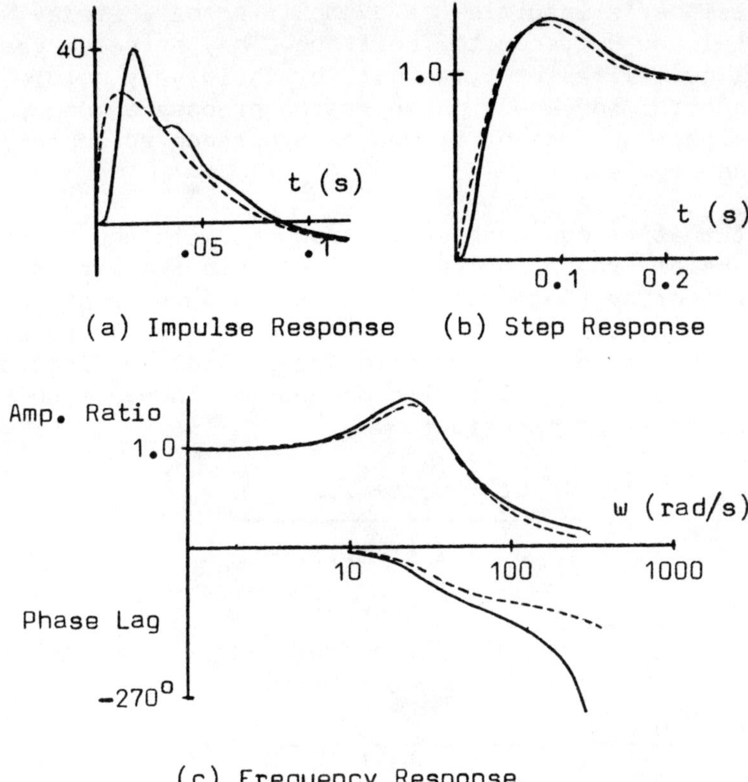

Fig (3.6): Dynamic and frequency response error of a low-order model of an electro-hydraulic servo. Actual system curves shown solid.

3.6. MODELLING USING STANDARD FORMS

Modelling has been considered so far in its most obvious context, namely the representation of an already existing system by means of a set of differential equations of reduced order. There is an alternative application however in which a model is selected as a design aim. That is, the model embodies the essential characteristics which are required of the final system and the design process is concerned with reproducing these features by suitable compensation of the existing hardware plant. Such models are known as standard forms and have been used extensively as analytic expressions of the designer's intuition as exemplified by Whiteley, Binomial and Butterworth functions. They serve as very useful guidelines at times but, by their very nature, are constraining when in the design process a number of specifications must be traded-off to reach an acceptable compromise.

An alternative approach was introduced with the investigations of Graham and Lathrop (24) into the use of the ITAE criterion (Integral of Time x Absolute Error) as a means of selecting optimum system functions. To reduce their results to standard form for general application they defined a normalization procedure. Thus, given the following system function:

$$T(s) = \frac{b_0 + b_1 s + \dots + b_m s^m}{a_0 + a_1 s + \dots + a_n s^n}$$

Now introduce a normalization frequency w_o such that:

$$w_o^n = \frac{a_0}{a_n} ; \; q_i = \frac{a_i}{w_o^{n-i} a_n} ; \; p_i = \frac{b_i}{w_o^{n-i} a_n} ; \; i = 1, \dots, n$$

The system function T(s) can now be rewritten in the form:

$$T(s) = \frac{p_o + p_1(\frac{s}{w_o}) + \dots + p_m(\frac{s}{w_o})^m}{1 + q_1(\frac{s}{w_o}) + \dots + q_{n-1}(\frac{s}{w_o})^{n-1} + (\frac{s}{w_o})^n} \qquad 3.21$$

The normalization is equivalent to a linear change in the time and frequency scales which enabled the authors to represent the responses of third-order systems in a two-dimensional form. However, it remained for Towill (25,26) to point out the potential of the technique for system design. Thus, although the unit step response is independent of w_o, other criteria in general use (such as bandwidth) may be represented by:

$$X = w_o^N . f(b,c) \qquad 3.22$$

where N is a positive or negative integer and b,c are (for a third-order model) equal to q_2, q_1.

The important point now is that all the various design criteria of interest can be plotted as curves on the b,c plane (known as the coefficient plane) with only the term w_o^N required to fix the time and frequency scales (27). The advantage of the model is that it makes immediately apparent the trade-off in performance which occurs as the location of the model within the coefficient plane is varied.

A coefficient-plane model which has found wide usage is defined (7,28):

$$T(\frac{s}{w_o}) = \frac{1 + ac(\frac{s}{w_o})}{1 + c(\frac{s}{w_o}) + b(\frac{s}{w_o})^2 + (\frac{s}{w_o})^3} \qquad 3.23$$

where with a = 1, T(s) represents a Type II system and with a = 0, T(s) is of Type I.

To illustrate the use of this model, say with a = 1, consider the situation in which a model is required with the following characteristics:

Step Response Peak Overshoot = M_p = 25%

System Bandwidth = w_B = 20 rad/s.

These criteria with others have been computed (27) as functions of the b,c co-ordinates of the coefficient-plane and are reproduced in figure (3.7). The shaded area of the figure represents the unstable region for a third-order system. On traversing the 25% overshoot contour it is seen that a range of values for the bandwidth can be selected to satisfy the specifications and the question now is which to choose? Had there been other specified criteria then they could have been used to define a unique location. As it is, with no other constraints, the most sensible choice would either be the point P, since here M_p is least sensitive to changes in the b co-ordinate, or the point Q where sensitivity to changes in c is least. The relationship of b and c to the parameters of the system would provide information on which of the two co-ordinates was most prone to variation.

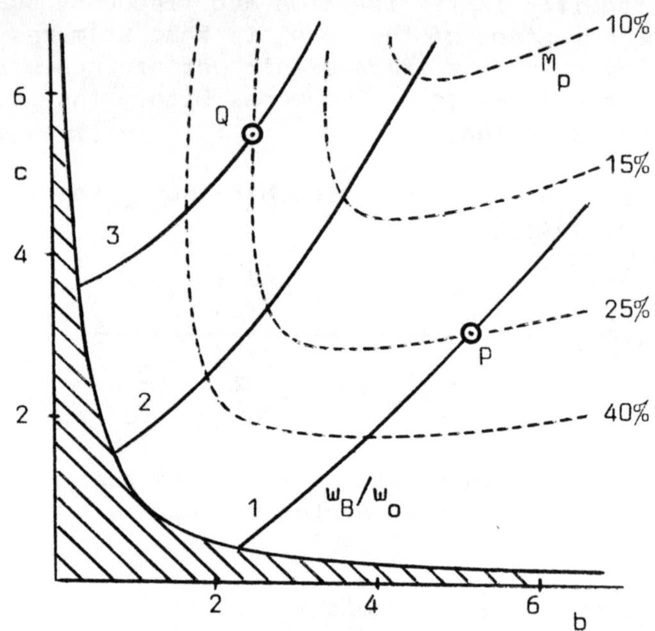

Fig (3.7): Peak overshoot M_p and normalized bandwidth w_B/w_o contours as functions of the coefficient plane co-ordinates b and c.

Suppose the point P is selected, then:

$$w_o = w_B = 20 \ ; \ b = 5.0 \ ; \ c = 3.1$$

The required model is therefore:

$$T_m(s) = \frac{1 + 0.155s}{1 + 0.155s + 0.0125s^2 + 0.000125s^3}$$

With other contours available, the designer can very quickly and with the minimum of effort arrive at some suitable model embodying the inevitable compromises which he can now make by inspection. A detailed account of the technique is presented in reference (7).

3.7. MODELLING FROM THE TIME RESPONSE

This is essentially the identification problem where the investigator arrives at or decides upon the structure and order of the model and then proceeds to estimate the values of the coefficients of the model. Structure and parametric values are usually arrived at by the iterative use of a computer algorithm involving optimisation between model order and the accuracy of the prediction. Common techniques include the recursive least-squares or generalized least-squares and the maximum likelihood methods. As these techniques lie beyond the scope of the text no further consideration will be given to them, but interested readers will find a good introduction to the subject in the work of Box and Jenkins (29). It may be apparent however from what little has been said that if the reduction of computing facilities is the main reason for low-order modelling then this is not the approach to adopt.

An alternative time-domain method is known as the method of moments (30) in which the transfer function is to be expanded as a polynomial in s with coefficients which are the successive time moments of the impulse response. Thus, by definition:

$$T(s) = \int_0^\infty e^{-st} h(t)\,dt$$

where h(t) is the impulse response.

If the exponential is now expanded, we have:

$$T(s) = \sum_{i=0}^{\infty} \frac{s^i}{i!} \int_0^\infty t^i h(t)\,dt \qquad 3.24$$

$$\text{where } M_i = \int_0^{\infty} t^i h(t)dt \text{ is the } i^{th} \text{ moment.}$$

The technique is to expand both the original and model transfer functions and then to match terms, where the number of moments required is equal to the number of parameters of the model. It is of course necessary to decide in advance the form of the model to be used and to derive the full transfer function of the high-order system. It has been pointed out (10) that as the order of the model increases considerable difficulty is experienced with convergence of the impulse moments for realistic test data and, even for convergent cases, there is inadequate resolution in estimation of the coefficients. Nevertheless, it is a powerful technique for use with heavily-damped systems.

3.8. SUMMARY

In this chapter an attempt has been made to introduce the reader to a field of great importance to the practising engineer by way of discussion of a number of the more commonly employed techniques. He should never dedicate himself to one particular method but rather to bear in mind their individual characteristics that he might, when involved with some aspect of performance, be in a position to select the most appropriate method or combination of methods to achieve his goal. Questions he should ask are:

Is reliability of prediction over a wide range of systems essential.

Should model order be obvious.

Are non-minimum phase terms a disadvantage.

Is the prediction of initial response, dynamic response and/or steady-state response required.

Is it essential for the simultaneous prediction of both transient and frequency response.

Is it necessary to model the impulse response.

Should there be a direct relationship between the model and the parameters of the original system.

To assist the reader, a resume is now presented of the advantages and disadvantages of the methods which have been discussed.

3.8.1. Loop-Function Method

Advantages : Simple; order clearly defined; no computational facilities required; very good prediction possible.
Disadvantages: Full transfer function required; Must employ (say) least-squares if only the measured data is available or if there are many terms in the cross-over region.

3.8.2. System Transfer-Function Methods

(1) Matsubara:

Advantages : Simple; No computational facilities are required; very good prediction possible.
Disadvantages: Model order not defined; full transfer function required; model involves an exponential term which may or may not present problems.

(2) Continued-Fraction:

Advantages : Easy to implement on computer.
Disadvantages: Can lead to unstable models or unsatisfactory initial time response prediction; no indication of model order; numerator order dictated by denominator; can have poor frequency response prediction.

(3) CF Method with Dead-Time:

This facility enables non-minimum phase terms to be eliminated at a cost of some distortion of response. It does however preserve the rank of the original system function and allows adjustment of model response to fit the measured data.

3.8.3. Least-Squares Frequency Response Method

Advantages : Does not require the full transfer function; Employs experimental data; Good prediction with adequate model orders.

Disadvantages: Model orders not defined; computational facilities are required; Can give rise to non-minimum phase models unless the model structure is constrained.

3.8.4. Standard Forms

Advantages : Simple to use; tabulated responses.

Disadvantages: Unduly constraining as a design tool.

(1) Coefficient-Plane Method:

Advantages : Simple; rapid; allows trade-off between specifications; good predictor for a wide range of systems; synthesis tool.

Disadvantages: Restricted to third-order models;requires data tables.

3.8.5. Time-Response Methods

Advantages : Does not require the transfer function if identification methods are used; Algorithms available.

Disadvantages: Extensive computation; full impulse response required for method of moments; model order not defined; possible problems with convergence.

CHAPTER 4

Plant Variation and System Sensitivity

4.1. INTRODUCTION

It has been suggested elsewhere that any difficulties which may be encountered in the attempt to achieve a specified system performance under closed-loop control arise as a consequence of the physical inability to apply sufficient control effort where it is required or from the lack of information on the dynamical changes which are occurring. That is, if the designer has total authority over the arrangement of power inputs applied and output measurements observed, then even the formulation of the feedback problem is under his control.

Be that as it may, the implicit assumption of such a proposition is that the designer has knowledge in the fullest sense concerning the nature of the system he is hoping to control. It is in the refutation of the latter statement that this chapter is devoted and, further, in the suggestion that ignorance of the system and its environment is the raison d'etre of feedback control theory.

However, does not our previous discussion of system models imply that everything of consequence is known about the system? Yes, provided we appreciate that by 'system' we mean a working hypothesis which should be a satisfactory representation of the theoretical plant under some specified conditions which will usually imply linearity, parameter-invariance and an operation free from those externally generated disturbances. It is suggested that what has been arrived at is a nominal model and that when the real system is in operation with production line hardware and is affected by changing environments the performance may be somewhat different to that which was expected.

It is primarily with these latter effects in mind that the feedback configuration will be decided. Sensitivity of the plant to component variability, parameter non-linearity and unforeseen operating conditions is the very basis out of which control engineering has grown. In an ideal world constant-coefficient differential equations would abound, but the practising engineer must never languish under such an illusion since the causes of variability are many and their consequences can be disastrous at worst or expensive at best.

4.2. PLANT IGNORANCE

Before confidence can be placed in the set of equations representing the plant we must have assured ourselves that the nature of the physical processes involved are fully understood and that their relationship to the external world is clearly defined. This in itself can be an immense problem but the designer must at least approximate to this degree of knowledge if he is ever to be in a position to distinguish with confidence between primary and secondary effects. That is, for him to be aware of what must be accounted for in the plant representation and what can be ignored (but borne in mind) as of second-order significance. This differentiation will sometimes be possible by experimental deduction. Indeed, with some complex plants whose physics are inadequately known, experimental methods may represent the only reliable route to a mathematical model.

Even now we can see that a measure of 'ignorance' will be built into the model which must ultimately be quantified if the full span of behavioural response is to be predicted. It should be apparent that the performance of the closed-loop system will be sensitive to the variation of the model as the physical conditions of the plant change. What other considerations must be taken into account for an estimation to be made of the variability of the plant? Some of the more common causes are presented below:

4.2.1. Ignorance of the physical process

This may arise because the plant was modelled from experimental data obtained from tests which did not stimulate all behavioural modes or because, as sometimes happens in thermodynamical processes, the parameters are known only vaguely. Ignorance may of course be actor deliberate as in the neglect of what are believed to be second-order effects.

4.2.2. Ignorance of the environment

No physical process can be dissociated entirely from its environment and will therefore respond to or be 'disturbed' by its vagaries as typified by tracking antennae in a gusting wind. Most natural disturbances can only be described in terms of probability distributions and therefore cannot be included in deterministic descriptions. We will return to such phenomena later.

4.2.3. Manufacturing tolerance

Every manufactured component carries with it a tolorance on its nominal value where it is normally assumed that the distribution of values about this mean value follows the Gaussian law with the spread, as defined by the tolerance, representing a three standard-deviation boundary. Compound this variability with that of the other components constituting the plant and it will be obvious that only a probabilistic model will suffice to express the variation of the system.

4.2.4. System nonlinearities

It is convenient in analysis and synthesis to assume that the process in hand can be described by sets of linear equations when in fact this is a fiction since all processes are nonlinear to some degree under some circumstances. A linear model is obtained only by linearization about one specific operating condition and will therefore vary whenever the operation of the plant changes.

4.2.5. Parametric variation

It is well known that components age with the passage of time when, as a result perhaps of wear and tear or just inherent deterioration, the system parameters which are determined by that component will change in either a catastrophic or asymptotic manner to a new value. Such long-term variation is of importance for the integrity of testing and tuning procedures which must be valid throughout the life of the system. If the plant exhibits time or cyclical variation, this behaviour will have to be reflected in the set of responses expected of the final system. A further source of parametric variation is observed with marine vessels the equations of which, besides being nonlinear, have hydrodynamic coefficients which vary according to the instantaneous motion of the vessel.

It is seen therefore that the mathematical model of any engineering system necessarily involves a number of approximations brought about by the simplification of configuration, reduction of order, elimination of non-linearities and the assumption of parametric invariance, each of which introduces a degree of uncertainty into any prediction of the performance of the closed-loop system. However, the designer must be able to express quantitatively, at the design stage, the degree of correspondence between the predicted and true behaviour of the system since the a posteriori modification by experience is embarassing, time consuming and expensive.

The remainder of this chapter is devoted to the sensitivity analysis of systems when the variations themselves can be considered as small perturbations. Although this constraint may appear to impose a severe limit to the usefulness of the techniques, it has been shown (14) to provide accurate predictions in practical situations.

4.3. THE SENSITIVITY FUNCTION

The sensitivity function is introduced as a means of deducing the variation which occurs in one dependent variable as a consequence of the changes occurring in some other independent variables. For instance if, for any reason, the transfer function of a plant P(s) should change, then the transfer function of the closed-loop system T(s) will also change by an amount which will be dependent upon the total loop function L(s) since:

$$T(s) = \frac{G(s)P(s)}{1 + G(s)H(s)P(s)} = \frac{G(s)P(s)}{1 + L(s)} \qquad 4.1$$

where G(s) is the forward-path shaping function and H(s) is the feedback function.

Let us assume that the plant changes by an amount δP and by expanding the resulting system function as a Taylor's series leads us to:

$$T(s,P+\delta P) = T(s,P) + \frac{\partial T(s)}{\partial P(s)}.\delta P + \ldots\ldots \qquad 4.2$$

Now on the assumption that δP represents only an incremental change, the higher-order terms of equation (4.2) may be discarded as of little significance and we have:

$$\delta T = \frac{\partial T(s)}{\partial P(s)}.\delta P$$

By re-arranging:

$$\frac{\delta T}{T} = \frac{\partial T/T}{\partial P/P}.\frac{\delta P}{P} = S_P^T.\frac{\delta P}{P} \qquad 4.3$$

The ratio of the proportional changes of T(s) and P(s) of equation (4.3) is defined as the classical sensitivity function S:

$$\boxed{S_P^T(s) \triangleq \frac{\partial T(s)/T(s)}{\partial P(s)/P(s)}} \qquad 4.4$$

Of course P(s) is a function of perhaps a number of variable quantities k_i, to each of which can be associated a plant sensitivity function:

$$S_{k_i}^{P}(s) = \frac{\partial P(s)/P(s)}{\partial k_i/k_i} \qquad 4.5$$

Simple manipulation now leads to:

$$S_{k_i}^{T}(s) = S_{k_i}^{P}(s).S_{P}^{T}(s) \qquad 4.6$$

If now S_P^T of equation (4) is evaluated using equation (4.1) for the general feedback structure, we find:

$$S_P^T(s) = \frac{1}{1 + L(s)} \qquad 4.7$$

It is seen from the beautifully simple expression of equation (4.7) that the sensitivity of the system function to changes of the plant is dependent only upon the loop function and that if the magnitude of the denominator of the expression is always greater than unity then the function of feedback has been fulfilled and the closed-loop system will be less sensitive to changes in the plant than the plant itself. If this is not the case however, the closed-loop system will fare worse than the open-loop plant - hardly a recommendation for the feedback configuration employed. And yet such a situation obtains when simple unity-feedback control is employed. Consider for example the typical loop function which is obtained with unity feedback when the phase margin is used as a guide to satisfactory transient response, as illustrated in figure (4.1).
A significant frequency range of the loop response is seen to lie within the circular boundary, implying that the magnitude of (1+L) is less than unity and that the closed-loop system will be inferior to the plant with respect to sensitivity within this range. This effect was given prominence by Bode (31) who showed that for any practical system having a loop function which

decayed to zero at infinite frequency at greater than 6 db/octave the following identity must hold:

$$\int_0^\infty \log_{10}(S_p^T)dw = 0 \qquad 4.8$$

That is, with such a system, there must be as much positive feedback as there is negative feedback.

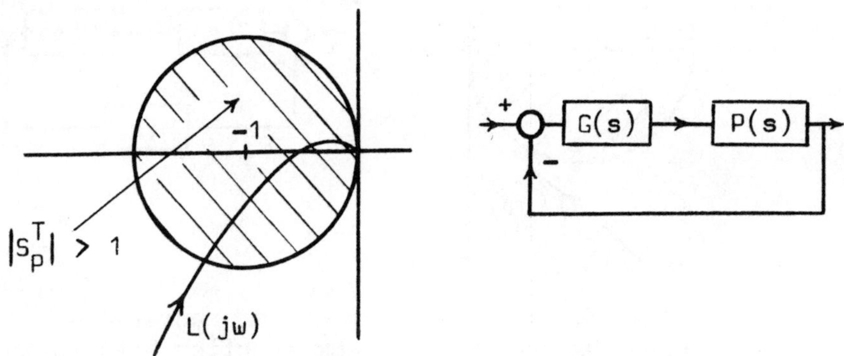

Fig (4.1): The sensitivity boundary of the loop-function plane consisting of a circle of unit radius centred on the Nyquist point and within which the closed-loop sensitivity is inferior to that of the open loop.

It is usually the case that control systems designed on the basis of T(s) alone, as in figure (4.1), result in systems which are extremely sensitive to plant variations at a critical part of the frequency response which is dominant in deciding dynamic performance. Thus, with the system above, should there be considerable plant variation, both resonant frequency and peak resonance will have exaggerated uncertainty. Such systems are said to have one degree-of-freedom since only either the system function T(s), which decides the dynamic response, or the loop function L(s), which decides sensitivity performance, is independent. Having decided one, the other is fixed.
However, by incorporating a feedback function H(s), the loop function can be chosen to ensure adequate sensit-

ivity whilst the forward-path function G(s) provides the freedom to select the required dynamic performance. For this reason such a system is said to have two degrees-of-freedom. An example of the improvement which is now possible with respect to sensitivity is shown in figure (4.2).

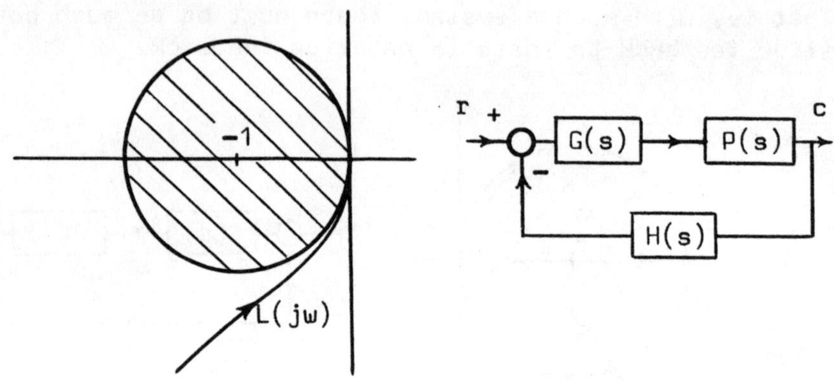

Fig (4.2): Exploitation of the second degree-of-freedom provided by the feedback function H(s) to ensure that the sensitivity function is less than unity over all frequencies.

The importance of the sensitivity function as a design criterion is emphasized when the response of a system to an external disturbance is considered. Thus, if the disturbance is assumed to enter the loop at the input to the plant of figure (4.2) and it is denoted by d, the transfer function relating the system output c(t) to d(t) is readily shown to be:

$$\frac{c}{d}(s) = \frac{P(s)}{1 + L(s)} = S_p^T(s).P(s) \qquad 4.9$$

Thus the sensitivity function is also a measure of how well a given feedback configuration will perform in the rejection of disturbances. It is therefore essential, for good system design, to give as much weight to the synthesis of the sensitivity function as it is to the system function.

4.4. THE COST OF SENSITIVITY REDUCTION

Consider the frequency responses of the closed-loop system T(jw), the plant P(jw) and the loop function L(jw) as shown in figure (4.3) for some hypothetical system.

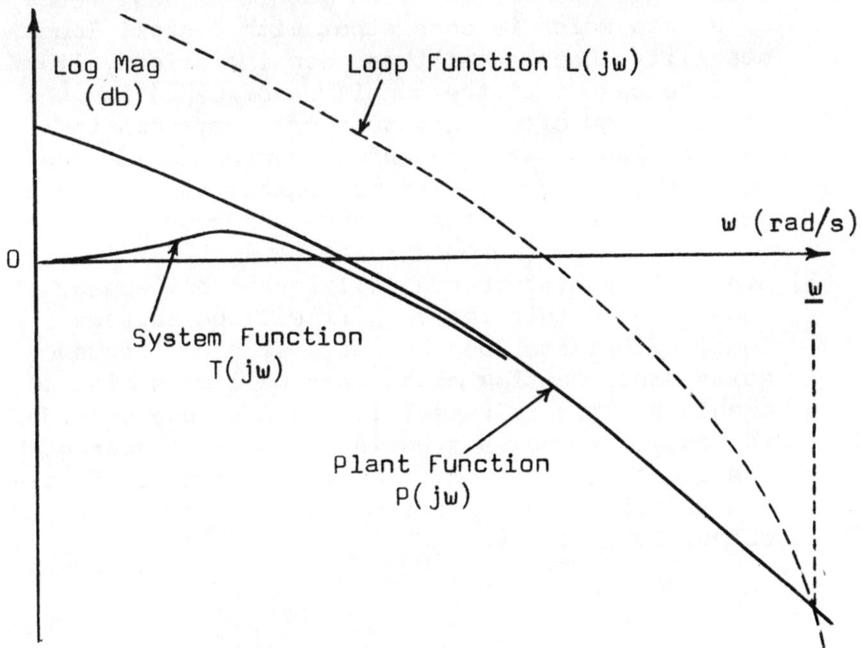

Fig (4.3): Illustration of the requirement for a large loop-bandwidth to achieve adequate reduction in sensitivity at the lower frequencies.

There are several points to note:

(1) At every frequency, the magnitude of the plant response is greater than that of the closed-loop system. This must be so since the physical capability of the plant must be at least equal to that required of the system. If more is demanded of the system than the plant can deliver then inevitable degradation of the performance will ensue as a result of saturation.

(2) Although the operational frequency range of the system may be only a few percent of the bandwidth

it is necessary to maintain authority over the frequency range up to say twice the bandwidth to ensure the desired transient performance. It will be necessary therefore to have high values of the loop frequency response over this range in order to achieve adequate sensitivity reduction. Beyond this range however the loop magnitude must decay at a rate which is consistent with overall loop stability. Because of these considerations, it will be seen that the magnitude of $L(jw)$ is greater, and often very much greater, than that of the plant over a frequency range $\underline{w}$ which can be considerably more extensive than the dominant frequency range of the closed-loop system.

(3) Since noise disturbance will invariably appear at some point within the loop it will be obvious that, with large loop bandwidths, high frequency noise amplification might very well give rise to problems. This is especially so when one considers the amplified noise signals which can appear at the input to the plant. Thus, the transfer function relating plant input θ_p to noise n on the output is given by:

$$\frac{\theta_p}{n} = \frac{G(s)H(s)}{1 + L(s)} = \frac{1}{P(s)} \cdot \frac{L(s)}{1 + L(s)} \qquad 4.10$$

Considering only the high-frequency noise where the loop function, although having a magnitude less than unity, is still greater than that of the plant, we see that the expression of equation (4.10) approximates to:

$$\frac{\theta_p}{n}(s) \simeq \frac{L(s)}{P(s)} \qquad 4.11$$

Examination of figure (4.3) highlights the point that, for low sensitivity, the magnitude of $L(jw)$ must be much greater than that of $P(jw)$ and hence that the noise amplication predicted by equation

(4.11) may attain high values. The possibility therefore exists for the plant to be operated in a condition of saturation due to even small levels of output noise disturbance.

It will now be seen that a call for the benefits of feedback brings with it some very obvious costs. First of all, if the loop function is to be defined up to the frequency $\underline{w}$ then devices will have to be incorporated into the loop which are capable of operating at these high frequencies. Secondly, in order to define $L(jw)$ at these frequencies it will be necessary to have definite knowledge of the plant over the same range. Such information is not always available. Finally, the plant signals which are to be fed back must be measured using some transducer which will invariably introduce its own noise contamination. With high loop gains over a broad band of frequencies the noise will be amplified and give rise to plant saturation problems.
The true cost of feedback is thus seen to lie in, and the consequences of, the maintenance of high loop gains over extensive frequency ranges (32, pp 101,280,360).

4.5. SAMPLED-DATA CONSIDERATIONS

It will be recalled from the theory of sampled-data systems that if ideal sampling were possible, the frequency spectrum $E^*(jw)$ of the sampled signal would be a periodic version of the spectrum of the continuous signal $E(jw)$ as follows:

$$E^*(jw) = \frac{1}{T_s} \sum_{m=-\infty}^{\infty} E\{j(w + mw_s)\} \qquad 4.12$$

where T_s is the sampling period and w_s is the sampling frequency.

Physically, the periodic property implies that the sampler is unable to distinguish between a signal of frequency w and one of frequency $(w+nw_s)$, where n is any integer. However, every practical system will employ a holding device which effectively filters the sampled signal, resulting in the attenuation of high frequency components as illustrated in figure (4.4).

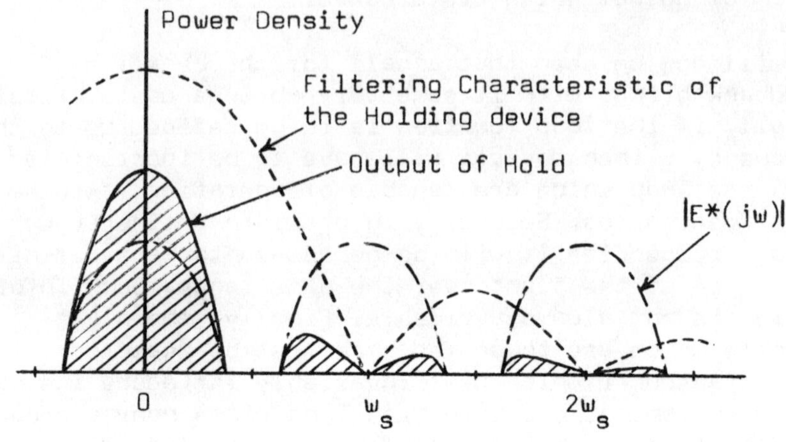

Fig (4.4): Frequency spectra of the sampled signal E*(jw) and the output of the hold-circuit.

It will be seen that the piece-wise constant signal which is applied to the plant has a spectrum which reproduces that of the original signal with the addition of the remnants of the sidebands caused by the sampling process. As long as the highest frequency component of the original signal is less than half the sampling frequency there is no further distortion. However, should this not be the case, then the original spectrum and the sideband spectra will overlap and a form of distortion known as aliassing will result. The spectrum of the signal to be sampled therefore must never be allowed to extend beyond $w_s/2$. As far as sensitivity is concerned this is a serious limitation since as we have seen the loop function bandwidth can be quite extensive. The decisive factor here is the sampling period which sets sets a boundary to the sensitivity reduction which can be achieved.

As serious as the limitation on the attainable loop bandwidth is when considered from the signal processing point of view, the situation is worsened when the effect of sampling is considered in the closed-loop context (33-36). It is well known that if the continuous plant P(s) has a pole-zero excess of at least unity then the combination of zero-order hold and plant, P(W), will have non-minimum phase zeros at W = 1 under the transformations:

$$z \triangleq e^{sT} \quad ; \quad W \triangleq \frac{z - 1}{z + 1} \qquad 4.13$$

where T is the sampling period.

Thus, although the continuous plant may be of minimum phase, the very act of sampling and holding converts it to one of non-minimum phase with the limitation on the loop gain which this implies. Thus, considering as an example the simplest of loop functions L(s) = 1/s, we have:

$$L'(z) = (1 - z^{-1})\mathcal{L}\frac{1}{s^2} = \frac{T}{z - 1}$$

$$L'(W) = \frac{T(1 - W)}{2W}$$

By representing the variable W as W = u+jv, where v is a fictitious frequency related to real frequency by the relationship v = tan(wT/2), we can draw the Bode diagram for the loop function as shown in figure (4.5).

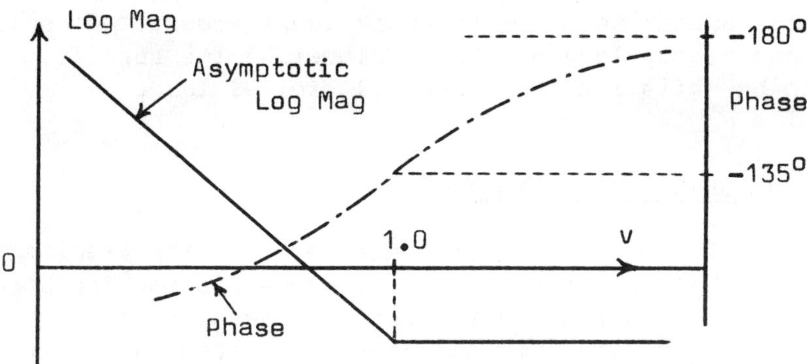

Fig (4.5): Illustration of the effect of the NMP zero introduced by sampling in increasing the phase lag beyond the fictitious frequency v = 1.

The NMP zero introduces increasing phase lag beyond the frequency $v = 1$ which may therefore be taken as a rule-of-thumb guideline to the maximum loop bandwidth if a satisfactory degree of stability is to be maintained. Thus, by relating this value of v to the real-world frequency ω, we find that the approximate limit for the loop bandwidth is given by:

$$\boxed{\omega = \frac{\pi}{2T} = \omega_s/4} \qquad 4.14$$

For a given sampling period T, then, there is a limit to the magnitude of the loop function and hence to the sensitivity reduction which is possible or, alternatively, to achieve a desirable loop function may require a much higher sampling frequency than expected.

Many factors influence the final selection of the period of sampling including such items as the control performance required, the control algorithm to be used, the spectrum of disturbances, accuracy of computation and the computational burden to be placed upon the digital processor. However, if the cost and inconvenience of rapid sampling is prohibitive, and if the loop-magnitude requirements must be met irrespective of any other considerations, then there may be no alternative but to involve both digital and analogue feedback. The latter will appear as minor feedback loops around the continuous plant, leaving the remainder of the specifications to be satisfied by an overall digital loop.

4.6. SENSITIVITY ANALYSIS

Sensitivity analysis is concerned with the examination of the effectiveness of an already-existing feedback structure in meeting the design objectives or in its comparison with some competative alternative system. It is not a design tool other than in an iterative mode of application and is more suited to the setting of tolerance gates for production line and in-service testing schedules. Three approaches which have appeared in the technical literature will be considered.

4.6.1. Frequency-Response Sensitivity

We have already seen how, using equation 4.6, the variations of the system function can be related to the k_i variable parameters of the plant. To obtain the equivalent frequency response variation we simply replace the Laplacian operator s by jω.

$$S_{k_i}^{T}(j\omega) = \frac{S_{k_i}^{P}(j\omega)}{1 + L(j\omega)} \qquad 4.15$$

For example, consider a unity-feedback control system with a forward-path function (the plant) defined by:

$$P(s) = \frac{1}{s(1 + s\tau)}$$

where the nominal value of the time constant τ is 2.0s.

Assuming the plant gain remains constant, the sensitivity function of the system with respect to the variable time constant is:

$$S_{\tau}^{T}(s) = \frac{-2s^2}{1 + s + 2s^2} \qquad 4.16$$

where of course the nominal values have been inserted.

The amplitude-ratio frequency response for the expression (4.16) is shown in figure (4.6) which, as expected for a single degree-of-freedom system, indicates an increase in sensitivity for frequencies above 0.58 rad/s where the loop function enters the unit circle. Consider only the change in T(s) at the resonant frequency ω_n = 0.707 rad/s brought about by a proportional change in the time constant of 10%. We have:

$$\frac{\delta T}{T}(j\omega) = 0.1414 \angle{-90^{\circ}}$$

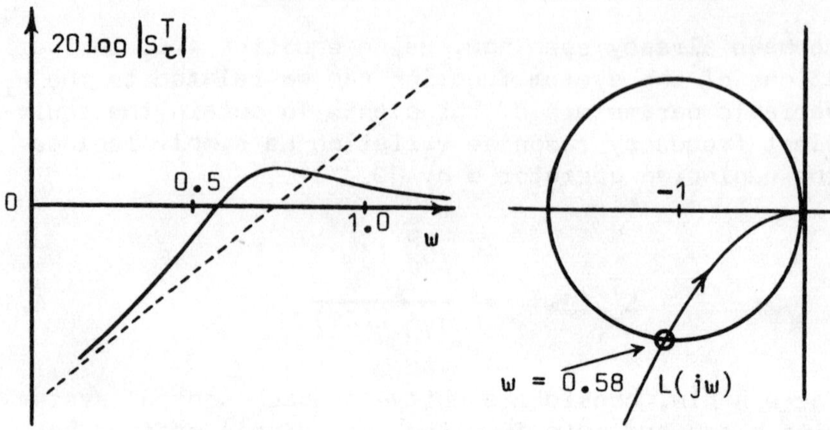

Fig (4.6): The sensitivity and loop functions for a unity-feedback control system with variable time constant.

On inserting the value of T(jw) at this frequency, we have:

$$\delta T = 0.2 \angle{-180^0}$$

The perturbed system function T' is therefore:

$$T' = 1.428 \angle{-98^0}$$

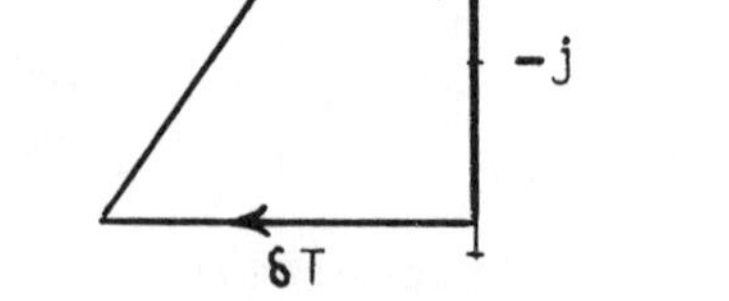

Note that, for this particularly simple case, although the 'incremental' change in T(jw) appears large, its effect on the system function is comparatively small because of their phase relationship. This is not always so. On calculating the true system function at this frequency and for the given change in the time constant we find $T(jw) = 1.4 \angle{-98.1^0}$ which compares well with that predicted by small perturbation analysis. However the user should always apply the technique with circumspection in situations where there is any likelihood of 'large' parametric variation or where the sensitivity function itself is large.

4.6.2. Sensitivity with respect to pole-zero migration

An alternative view of the sensitivity of the system function is obtained by examining its variation as the poles and zeros of the plant are allowed to vary. Although this aspect of sensitivity analysis is only a particular case of the parametric variation which has been discussed and although it has significance only for the dominant singularities of the system, the conclusions which can be drawn are of general interest in themselves and especially when using low-order system models (37).
Represent the plant P(s) as follows:

$$P(s) = \frac{K \prod_i (1 + \frac{s}{z_i})}{\prod_j (1 + \frac{s}{p_j})} \qquad 4.17$$

where z_i and p_j are the open-loop zeros and poles of the plant.

Now, assuming a forward-path compensator G(s) and a feedback function H(s), the closed-loop function can be written:

$$T(s) = \frac{KG \prod_i (1 + \frac{s}{z_i})}{\prod_j (1 + \frac{s}{p_j}) + KGH \prod_i (1 + \frac{s}{z_i})} \qquad 4.18$$

On partial differentiation of equation (4.18) with respect to the variable singularity q, where q is equal either to z_i or p_j, the sensitivity function is found to be

$$S_q^T = \frac{\pm s/q}{(1 + \frac{s}{q})} \cdot \frac{1}{1 + L(s)} \qquad 4.19$$

The sign of expression (4.19) is positive if $q = p_k$, or negative if $q = z_k$. One again notices the influence of the basic sensitivity function involving the loop response but to which is now added a factor which is indicative of the significance of the pole or zero throughout the frequency range. Thus, any change in the location of the singularity will have nil effect on the system frequency response at low frequencies, whereas at high frequencies the total plant variation will be reflected in the variation of the closed-loop system.

Sensitivity information presented in the above form is of a qualitative nature essentially and does not lend it self readily to quantitative evaluation or comparison. This is because pole-zero arrays are difficult to interpret in the time domain without recourse to further computation and very similar performances can be obtained from widely different configurations (26,38). These arguments are applicable to the alternative state-space representation where it is the sensitivity of the eigenvalues and eigenvectors which are of importance (39). Indeed it has been asserted (38) that state-variable sensitivity theory is 'noteworthy for its obscurity and confusion as compared with alternative frequency domain methods'.

4.6.3. Transient Response Sensitivity

Just as familiarity with the frequency response and its interpretation becomes second nature to the control engineer, so the designer quickly acquires an intuitive 'feel' for the implications of a particular form of transient response. If this insight is to be of any use however, he must also be aware at the design stage of just how much the nominal response will be modified as a consequence of inevitable parameter variation. With this in mind, a transient response sensitivity function S_k^u can be defined as follows:

$$S_k^{u(t)} \triangleq \frac{\partial u(t)}{\partial k/k} \qquad 4.20$$

where $u(t)$ is the unit step response of the system and k a system parameter.

Since u(s) is simply T(s)/s, simple manipulation leads to the following expression:

$$S_k^{u(t)} = \mathcal{L}^{-1}\left\{\frac{1}{s} \cdot T(s) \cdot S_p^T \cdot S_k^P\right\} \qquad 4.21$$

Equation (4.21) may be expressed more succinctly in terms of the implied convolution if it is noticed that (for unity-feedback systems) S_p^T is actually the transfer function relating system error to system input, thus:

$$S_k^{u(t)} = h(t)*\{1 - u(t)\}*S_k^P(t) \qquad 4.22$$

where h(t) is the impulse response.

An intuitive approach to the convolution of equation (4.22) has been reported in the literature (40) which besides handling the non-unity feedback case, leads to an insight into the advantages of feedback compensation in sensitivity reduction. The sensitivity function of equation (4.21) is computed most conveniently using an algorithm by Liou (41) or, alternatively, using an analogue machine to simulate the unit step response of the function $T(s)S_p^T S_k^P$. Note that, as with all sensitivity computation, the order of the functions involved is at least twice that of the system itself. In an attempt to reduce the computational requirement, investigations have been carried out (14) which verified that a well matched dominant low-order model is capable of very good prediction of the transient response sensitivity of high-order systems.

Let us now consider the application of the sensitivity function to the estimation of the changes occurring in the transient response of the system shown in figure (4.7) as a consequence of say a 50% change in the motor time constant T_m. The sensitivity function of the plant with respect to T_m is readily shown to be:

$$S_{T_m}^P = \frac{-sT_m}{1 + sT_m}$$

... which leads to the expression,

$$S_{T_m}^{u(t)} = \mathcal{L}^{-1} \frac{-4s}{(s^2 + 2s + 4)^2}$$

The latter function, evaluated for a nominal T_m of unity is illustrated in figure (4.7) together with the actual responses of the nominal and perturbed systems. To estimate the perturbed response, given that of the nominal system, the following expression is computed at the time instants of interest.

$$u^*(t) \cong u(t) + \delta u(t) = u(t) + S_{T_m}^{u(t)} \cdot \frac{\delta T_m}{T_m}$$

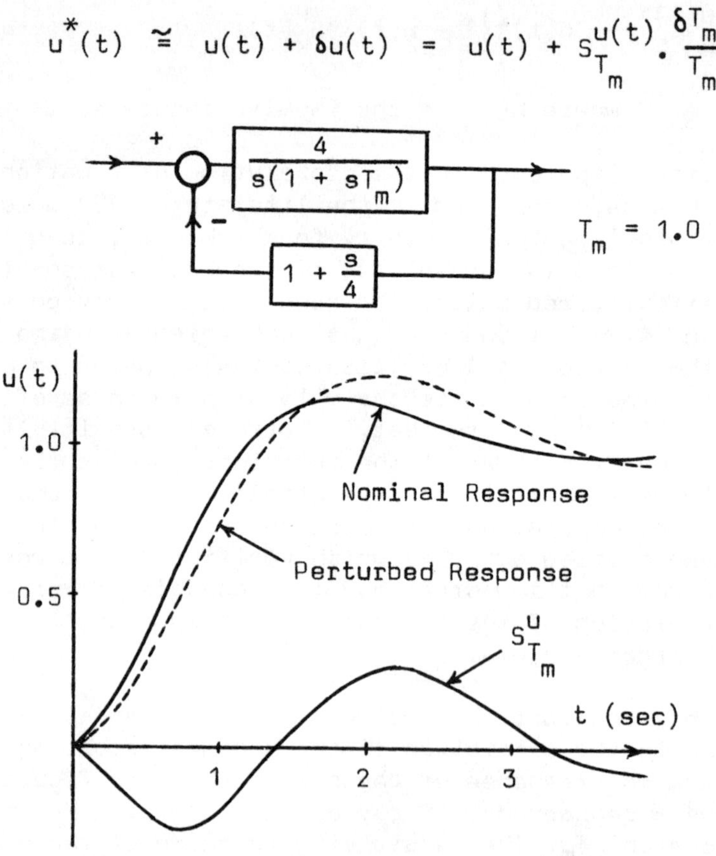

Fig (4.7): Comparison of the nominal and perturbed step responses with a 50% increase in the value of T_m. Also illustrated is the incremental transient response sensitivity function with respect to T_m.

Table (4.1) lists the true and estimated values of the perturbed response on the assumption of a 50% change in the motor time constant.

TABLE (4.1): Error introduced into the estimation on applying small perturbation analysis to a system with a parametric variation of 50%.

t	u*(t)	
	Actual	Estimated
0.5	0.256	0.230
1.0	0.730	0.729
1.5	1.098	1.106
2.0	1.241	1.280
2.5	1.204	1.195
3.0	1.091	1.047

In view of the fact that a 50% change in a parameter can in no way be regarded as a small perturbation, it is considered that the estimates of Table (4.1) are adequate for initial design purposes. The value of the transient response sensitivity functions is their visual impact and when combined with the insight provided by the technique already quoted (40) indicate the direction of further improvement of the loop function.

4.7. COMPONENT TOLERANCING

The previous section was concerned with the estimation of response variation for a fixed change in one particular parameter. However in practice one is confronted with a number of variable parameters which assume values anywhere within their possible range of variation as specified by their tolerances. Now the common assumption when reference is made to tolerance is that the probability distribution of the parameter about its mean value is Gaussian, or Normal, and that the limits expressed by the tolerance refer to the three standard-deviation boundaries of the distribution. That is,over

99% of all parameter values will be found to lie within the tolerance range. Not all parameters, if any, will take on their maximum deviation and so a means must be found which takes account of this and which provides the designer with a statistical boundary to the variation expected of the transient response. Now the total variation is made up of the sum of the individual variations induced by all the changing parameters. We can therefore write:

$$\delta u(t) = \delta u_1(t) + \delta u_2(t) + \ldots\ldots$$

where the δu_i are the contributions of the individual parameters.

Now, on taking expectations,

$$E[\delta u(t)] = E[\delta u_1(t)] + E[\delta u_2(t)] + \ldots.$$

Now on the assumption that the perturbations do have a Gaussian distribution with zero mean, the mean of the response variations will be zero and hence the variance can be expressed as:

$$E[\delta u(t)^2] = E[(\delta u_1(t) + \delta u_2(t) + \ldots.)^2]$$

$$= E[\delta u_1(t)^2] + E[\delta u_2(t)^2] + \ldots\ldots$$

$$\ldots + E[\delta u_1(t)\delta u_2(t)] + \ldots\ldots\ldots$$

Since the variations are assumed to be random, the expectation of terms formed by the product of unlike variations must be zero, which leaves the identity:

$$\sigma_u^2 = \sigma_{u_1}^2 + \sigma_{u_2}^2 + \ldots\ldots \qquad 4.23$$

where σ refers to the relevant standard deviation.

Finally, since the variations of each parameter are Normally distributed, the contributions of all parameters x_i to the variation of the transient response will also follow the same distribution, enabling us to write for the response deviation:

$$\Delta u(t)\Big|_{x_i} = 3\sigma_{u_i} = S^{u}_{x_i}(t) \cdot \frac{3\sigma_{x_i}}{x_i} \qquad 4.24$$

Thus, combining equations (4.23,24), the tolerance which can be attributed to the response at any time t is given by equation (4.25).

$$3\sigma_u(t) = \sqrt{\sum_i \left[3\sigma_{x_i} \cdot \frac{\partial u(t)}{\partial x_i}\right]^2} \qquad 4.25$$

The latter equation can be used in two ways. Firstly, it can be used in the direct sense for the computation of the response boundaries for a given set of toleranced components. Secondly, it can be employed as a means of deciding the tolerance which must be imposed upon the components which constitute the system if a given boundary envelope is to be satisfied. This is a most important role since too high a component tolerance will result in too high a probability that the dynamic response will be unacceptable, whilst an unnecessarily tight tolerance will result in excessively expensive components. Towill and Mehdi (14) consider the sixth-order system shown in figure (4.8) where the problem was to tolerance T_n, T_r and K for Normally distributed variations of the parameter T_a (nominal value = 0.5 and tolerance = $\pm$ 0.25).

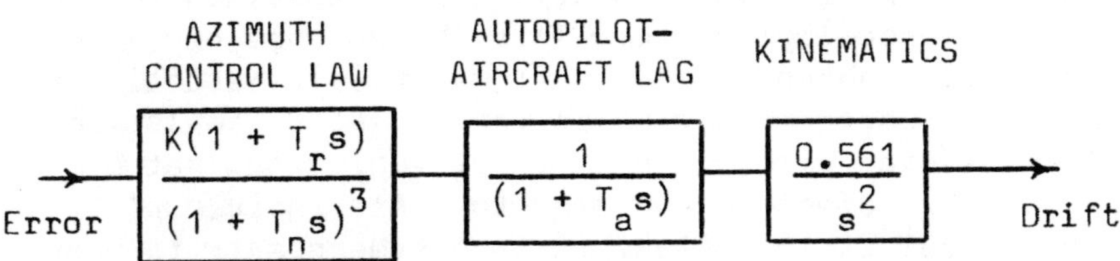

Fig (4.8): An aircraft azimuth-channel system.

Initial design indicated nominal values of K = 0.033, T_r = 10 and T_n = 0.73. Of course, before the estimation of the tolerances can be attempted, the designer must be aware of those boundaries of the transient response which correlate with unsatisfactory systems. This is in itself of great importance as when the designer reaches the point of the setting of testing specifications (42). However, in the present example, the authors relied upon response templates as illustrated in figure (4.9) based on an agreed interpretation of pilot opinion. From an examination of the sensitivity functions of each of the four parameters and using a trial-and-error procedure, the authors were able to specify suitable tolerances for each parameter as K (10%), T_n (25%) and T_r (25%).

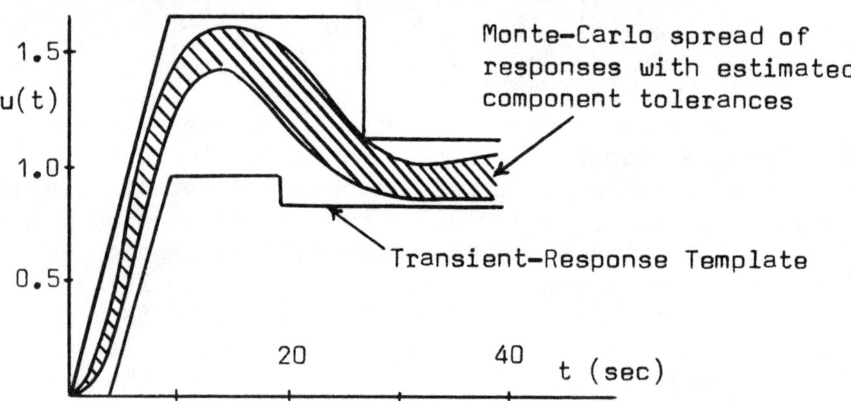

Fig (4.9): Specified template boundary for transient response and the spread of the observed responses for the system of figure (4.8).

Because the analysis has depended upon the assumption of small perturbations, it is prudent to validate the results by performing a Monte-Carlo simulation (43,44) of the closed-loop system. This the authors did with the results shown shaded in figure (4.9). Essentially, the Monte-Carlo method calls for repeated simulation of the system with, at each iteration, a new set of parameters selected according to their individual statistical distribution. It is not necessary to assume that each parameter under investigation has a Normal distribution since, at the beginning of each iteration a uniformly-distributed number is generated in a random manner which must be processed by standard means (44) to be a variate of the required form of distribution. Thus, with each new parameter set, the characteristics of the response are computed and stored until after a suitable number of iterations (certainly more than 100) the spread of the characteristics can be ascertained.

4.8. SUMMARY

The mathematical model of an engineering system necessarily involves a number of approximations requiring the designer to predict at the design stage their effect upon system performance. To this end the sensitivity function was introduced which related either the system frequency or transient response to changes in either the location of the system poles and zeros or, more usefully, the values of the system parameters. This function was seen to be equally applicable to sampled-data as well as to continuous systems for analytical purposes and in its important role for the setting of component tolerances.

The ultimate limitation of the technique lies in its initial formulation from the Taylor's expansion of the perturbed system function, requiring as it does that all changes should be incremental. The question then arises as to what constitutes a small change? However, the fact remains that for some systems the parameters may undergo very large variations (greater than 100%) which calls for an alternative design method. This will be considered in the next chapter.

CHAPTER 5

Loop Synthesis and Sensitivity Reduction

5.1. INTRODUCTION

It has been seen that the judicious use of feedback in the control of plants brings with it the attendant benefits of disturbance rejection and the reduction of the sensitivity of the closed-loop system to changes in the parameters of the plant, both of which were seen to be dependent upon a sensitivity function defined in terms of the loop function L(s) alone. This important function is repeated below.

$$S_p^T(s) = \frac{1}{1 + L(s)} \qquad 5.1$$

However, a serious short-coming of equation (5.1) is that, as defined in the previous chapter, it is applicable only to incremental changes of the parameters. This must of necessity imply some lack of confidence in the results of an analysis should the parameters undergo large or even moderate variations.

To overcome this difficulty, two methods are introduced in this chapter to facilitate the design of a feedback system to control, within specified bounds, a plant which can exhibit massive variation. If, further, the plant can be said to be of the minimum-phase category, it is possible to design the closed-loop system with any degree of insensitivity to such variations as may be required. The first method employs familiar root-locus techniques whilst the second is based upon the moulding or shaping of the loop frequency response to achieve the desired objectives. The former is applicable to systems which have responses dominated by only a few significant poles and zeros whereas, at the cost of a little more effort, the latter method can be applied universally.

5.2. ROOT-LOCUS SYNTHESIS

The reader will recall that the root locus technique is a graphical means for determining the migration of the closed-loop poles as a function of the loop gain by a suitable interpretation of the characteristic equation:

$$1 + L(s) = 0 \qquad 5.2$$

If $L(s)$ is written in terms of its numerator, $n(s)$, and denominator, $d(s)$, polynomials then equation (5.2) can be expressed as:

$$d(s) + Kn(s) = 0 \qquad 5.3$$

where K is the loop gain constant.

It is obvious that when K is zero the roots of equation (5.3) are the open-loop poles whilst, when K increases indefinately, the closed-loop poles approach the open-loop zeros. The loci of all possible closed-loop poles therefore begin at the open-loop poles and terminate on the open-loop zeros, implying as many loci as there are open-loop poles. Also implied is the fact that, since $L(s)$ is a rational algebraic function, there must be equal numbers of poles and zeros, some of which may be located at infinity. There will therefore be a number of loci which tend to infinity which is equal to the pole-zero excess of $L(s)$.

Some consequences of the foregoing are worthy of a brief comment. Firstly, it will be apparent that should the loop function possess zeros which lie in the right half-plane (RHP), known as non-minimum phase terms, then with sufficient loop gain poles of the closed-loop system will inevitably appear in the RHP with consequential system instability. In such cases therefore, there is an upper limit to the magnitude of the loop function which can be allowed and hence a limit to the sensitivity reduction which is possible. It will also be appreciated that, with minimum-phase plants, it is theoretically possible for the designer to introduce loop compensation and so position poles and zeros at will to give him a wide degree of freedom in the final positioning of the closed-loop poles. This is most useful in such cases as when the plant possesses RHP poles which can be induced into the LHP as stable closed-loop poles or to inhibit the de-stabilising effects of secondary modes lying in close proximity to the imaginary axis of the s-plane. It is in such instances that the usefulness of root-locus diagrams are manifest by enabling the designer a visual insight into what can and what cannot be achieved and what is the minimum compensation which will be necessary.

Finally, and more to the point of this chapter, is the effect of gain changes on the location of a closed-loop pole as it migrates towards a zero. It should be obvious that at high levels of loop gain the proportional change of pole location with respect to change of gain will be greatly reduced. That is, a given fractional change of gain from a nominal high gain will have much less effect on the location of a closed-loop pole than will the same fractional change about a lower loop gain. In other words, the presence of an anchoring zero in close proximity to a desired closed-loop pole location will have the effect of reducing the sensitivity of that pole to any changes occurring in the loop frequency response as a result of plant variation.

Consider for example the simple system of figure (5.1) when the gain K is allowed to assume increasing values. It is readily seen that the closed-loop poles are given by:

$$s = \frac{-(K+1) \pm \sqrt{(K-1)^2 - 4}}{2}$$

Thus, with sufficiently large K, the dominant closed-loop pole lies within an infinitesimal distance of s equal to −1 and that if K changes, whilst still remaining large, there is negligible change in the location of the pole.

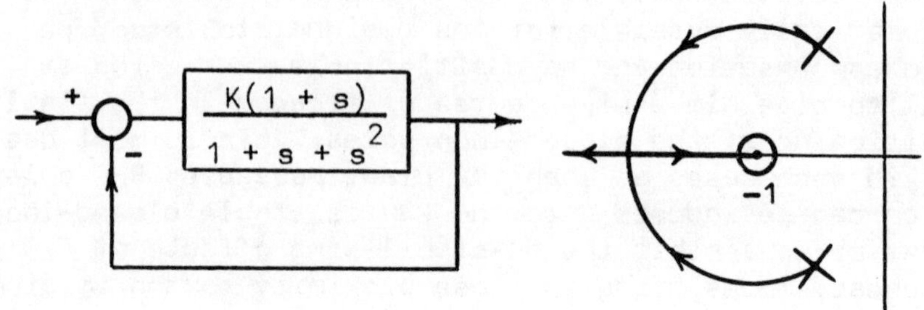

Fig (5.1): Block diagram and root-loci of a system used as an example of the desensitization effect of a proximate zero.

An alternative interpretation as we have seen is that closed-loop poles which lie in close proximity to zeros tend to be less sensitive to plant variations than would otherwise be the case. For example, with the case just considered, let two possible design locations be −1.27 and −1.01 and let the loop gain in each case be increased by 100%. It is found that the percentage change in location for the first choice is 12.6% whilst that for the second and closer choice is 0.5%. Of course these results should not be surprising since there is a 20:1 ratio of gains for the two cases and, naturally, the greater the loop gain the less the sensitivity function. Considering the proximity of poles to zeros is only an alternative viewpoint of the sensitivity scene.

5.3. SYNTHESIS VIA ZERO ASSIGNMENT

It is apparent from what has been said that an intuitive approach to system synthesis is to locate zeros in close proximity to the required closed-loop poles in the certain knowledge that, with sufficient zeros, the required response will be achieved with the stability guaranteed and with an in-built bias against the effects of parameter variation. An advantage of the approach is that the root-loci do not have to be plotted accurately but rather, the s-plane pole-zero array is used as an indication of alternative compensation schemes. It must be noted however that two assumptions are implied in the foregoing. Firstly, that the user is able to correlate a required transient response with a particular set of poles and zeros and, secondly, that he is in a position, physically, to provide the assigned zeros given the limitations in monitoring his hardware plant. The first difficulty fades into insignificance if the plant consists simply of a few dominant poles, whilst the second, if physically impossible due to a lack of measurement points, will involve the designer in an area of control theory known as state estimation which employs Observers and Optimal Filters (45,46). However, when the system is represented in state-variable format and complete state feedback is contemplated, the method has been formalized by Schultz and Melsa (47) as the H-Equivalent method.

The zeros already referred to are, of course, those of the loop function and the question arises as to their implementation once their number and locations have been decided. If some of them are required to appear in the closed-loop system transfer function then they must be introduced into the forward-path compensation. If this is the case then they must be accompanied by an equal number of forward-path poles to ensure physical realizability. Further, to prevent any modification of the desired response, they must be located remote from the region of dominant poles, that is, well beyond the system bandwidth. Care must be exercised at this atage if stability problems are not to be encountered. On the other hand, zeros may appear in the feedback function with or without associated poles since it is often possible to make transducer measurements of the derivatives of the plant output using tacho-generators, rate gyros and accelerometers. (But note that these devices have their own dynamics aside from the inherent differentiation and that if amplification is necessary the dynamics of the amplifiers must be taken into account also). For instance, if it were possible to measure the output, its rate and its acceleration and to feed back a weighted sum of these variables for the purposes of comparison, then a feedback function has been created which provides two loop zeros without any associated poles as shown in figure (5.2).

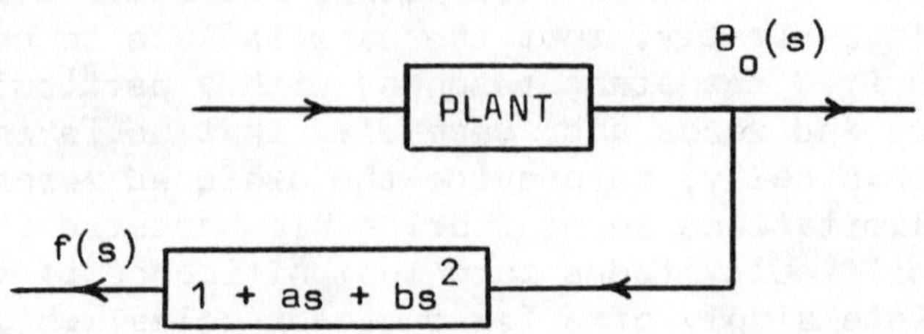

Fig (5.2): Introduction of two free zeros into the loop function by the expedient of additional rate and acceleration feedback:

$$f(t) = \theta_o(t) + a\dot{\theta}_o(t) + b\ddot{\theta}_o(t)$$

It is not always necessary to insist on the measurement of derivatives directly since often the filtering of available signals will suffice as instanced by the use of transient rate feedback (the approximate differentiation of a rate signal using a CR circuit). The consequence of this, however, is to introduce feedback poles which, unless cancelled by the forward path, will be found as closed-loop zeros.

Example:
A particular hydraulic motor has the transfer function given in figure (5.3) which, as is common with such devices, includes a very lightly damped resonance. It is required to modify the plant with minor-loop feedback prior to the overall design such that the damping ratio is 0.5, with a resonant frequency of 50 rad/s, without losing the Type I characteristic of the motor.

On drawing the pole-zero array it becomes immediately obvious that a quadratic zero-pair is the minimum compensation necessary to achieve the required poles and that to maintain the Type I characteristic the pole at the origin must be fixed in situ using another feedback zero. The two zeros could be obtained by providing two lag-lead networks in the forward path with the poles at some remote location. However, the figure shows two complex zeros provided by the feedback function. Such a third-order feedback function might be impossible to realise in practice and would call for the estimation of the required variables as discussed already. Be that as it may, we know what is required just by examination of the diagram and by a simple comparison of the required minor-loop function with the loop of figure (5.3) the unknown coefficients a and b can be found.

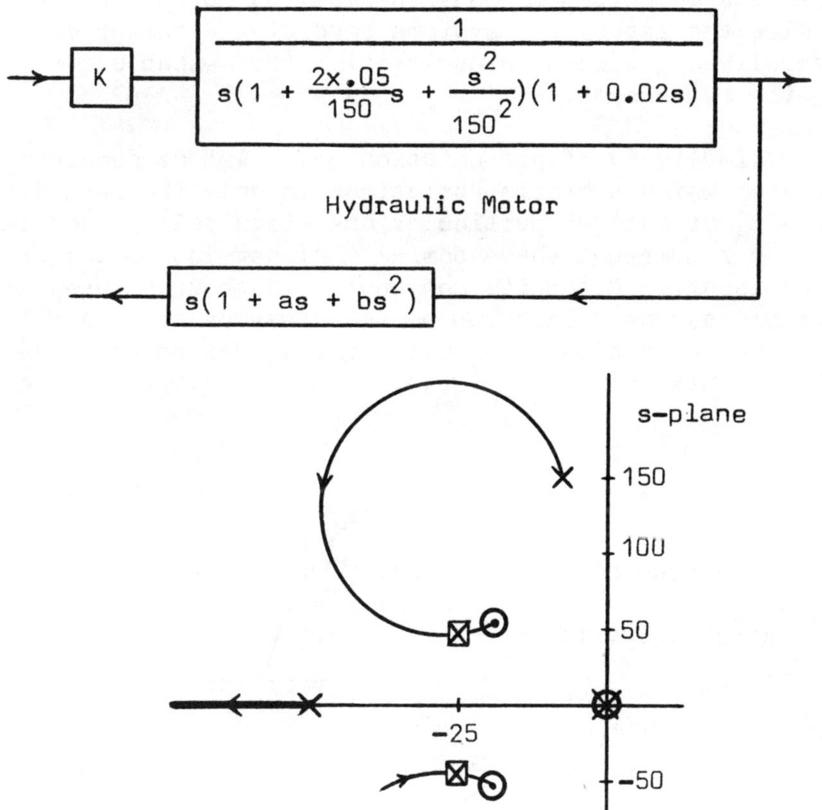

Fig (5.3): Block diagram and root-loci for the hydraulic motor compensated by minor-loop feedback considered in the text. The blocked-in poles indicate the required closed-loop pole locations. Direct comparison of the coefficients provides the gain K and the necessary zero locations.

5.4. SENSITIVITY REDUCTION VIA THE S-PLANE

The philosophy of the method is that loop zeros which are located near the desired dominant closed-loop poles (which we will call anchor zeros) in conjunction with sufficient loop gain will ensure that the roots of the characteristic equation will lie within a restricted region close to the zeros despite variation of the plant poles. It is important to note the use of the word dominant since the method is practicable only for systems which can be modelled well using only a few poles and zeros. For systems involving a number of equally significant singularities the method becomes quite cumbersome.

For simplicity of presentation only, let us consider a system which exhibits variations in only the gain K of the plant and one particular open-loop pole p_1 and let it be required to have dominant closed-loop poles at the location R and its conjugate. To achieve this, loop zeros have been inserted as anchor zeros at z_1 and its conjugate in close proximity to the desired poles as illustrated in figure (5.4).

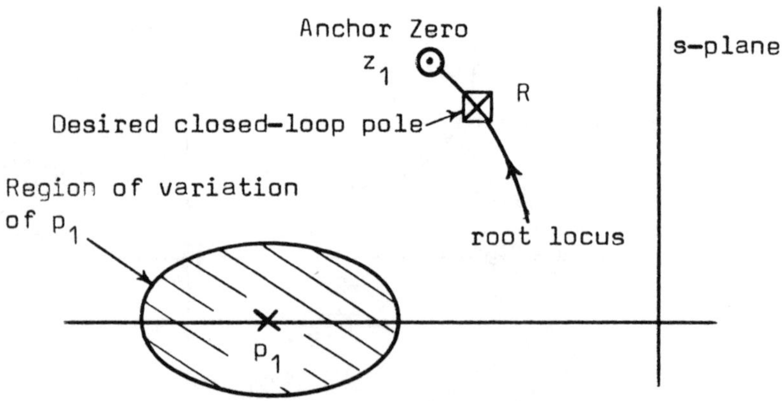

Fig (5.4): The anchorage of the desired closed-loop pole R within a restricted region by means of the inserted zero z_1. NB, except for p_1, the variable pole, all other poles and zeros have been omitted from the figure.

For the nominal values K_o and p_1, the magnitude criterion for the characteristic polynomial to ensure the location of the closed-loop pole gives:

$$K_o = \frac{\prod (p_i R)}{\prod (z_j R)} \qquad 5.4$$

where $p_i R$ and $z_j R$ represent the vectors from all poles and zeros to the point R.

If now the gain and variable pole change to K and p_1' respectively such that the closed-loop pole changes to a new location R', then:

$$K = \frac{p_1' R' \prod p_i R'}{z_1 R' \prod z_j R'} \qquad 5.5$$

Further, on the basis that R should not be permitted to move very much, the vectors from all the poles and zeros (except for z_1 which is close to R') will change very little. That is, $p_i R' \cong p_i R$ and $z_j R' \cong z_j R$. Thus we have:

$$K \cong \frac{p_1' R \prod p_i R}{z_1 R' \prod z_j R} \qquad 5.6$$

Thus, using equations (5.4,6), we obtain:

$$\frac{z_1 R'}{z_1 R} = \frac{K_o (p_1' R)}{K(p_1 R)} \qquad 5.7$$

If we know the range of variation in the s-plane of the pole p_1 and the change which can occur in K, then using equation (5.7) we can obtain the vectorial relationship between z_1R and z_1R' as the plant pole varies. Note that as yet we do not know the absolute location of the zero. However, we can plot the region within which R' will be located in terms of the unscaled vector z_1R as shown in figure (5.5).

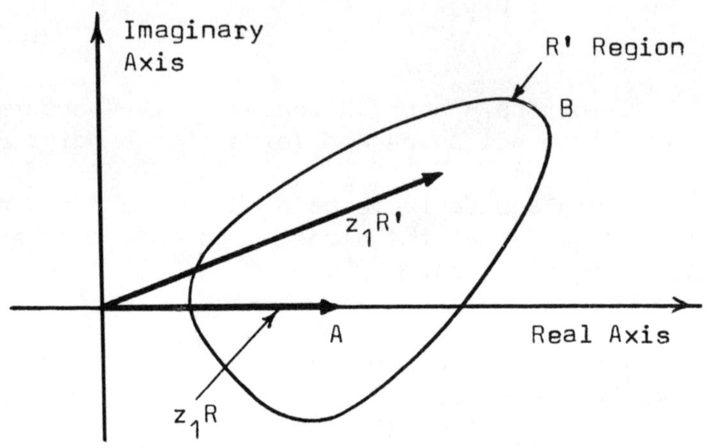

Fig (5.5): Relationship between the vectors z_1R and z_1R' as the gain and pole location change. The region is obtained from equation (5.7) of which the RHS is known from plant specifications.

To fix the scale of the vectors of figure (5.5), use is made of the tolerance specifications on the maximum allowable variation of R'. Thus, in the above figure, if R' must lie within a circle of radius 1.0 units about R, then the length AB (representing the maximum deviation of R' from R) must be 1.0, from which we can deduce the magnitude of the vector z_1R. That is, we now know the distance of the inserted anchor zero from the desired closed-loop pole. The only remaining information we need to know is the true angle of the vector z_1R. This is obtained from the knowledge that for R to be a closed-loop pole, the sum of the angles of the vectors from all open-loop poles and zeros (with due attention to sign) to R must, according to the phase criterion of the root-locus method, be an odd multiple of -180°.

As an example, consider the plant below for which z_2

lies at -2, p_1 = -1, p_2 = -4 and K is nominally unity. However each singularity can vary by 25% whilst the gain may increase to the value of 2. It is required that the dominant poles be located at $-1 \pm j$ but that as a result of the variation of the plant they may be permitted to lie within a circular boundary about the nominal location with a radius of 0.15. It is necessay to determine the feedback function on the assumption that all derivatives are available for feedback.

$$P(s) = \frac{K(s - z_2)}{s(s - p_1)(s - p_2)}$$

If derivative feedback is to be used as suggested then no extra poles will be involved and, as can be seen from figure (5.6), only two feedback zeros are required to fix the dominant complex poles.

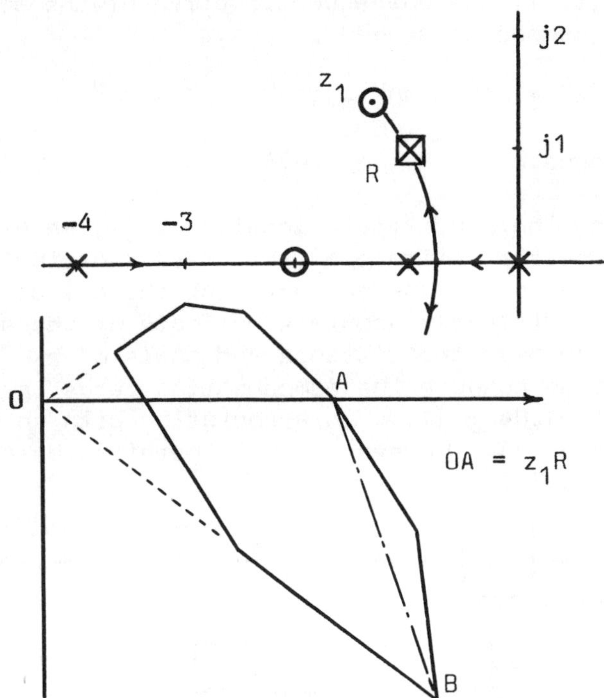

Fig (5.6): Tentative root-loci of compenated system and vectorial relationship between z_1R and z_1R' over the range of plant variation.

Initially we need to know the relationship between z_1R and z_1R' and this we find for the limiting locations of the singularities and the extremes of gain using:

$$\frac{z_1R'}{z_1R} = \frac{K_o}{K} \cdot \frac{(p_1'R)(p_2'R)(z_2R)}{(p_1R)(p_2R)(z_2'R)}$$

Note that the above expression differs from equation (5.7) since now three singularities are involved in the variation. On the assumption of 25% variation and taking a number of extreme case situations, the RHS of the above expression has been plotted in figure (5.6) from which it can be seen that AB represents the maximum variation of R' from R and that if this is to be 0.15 as specified, then z_1R = OA = 0.14. The angular relationship between z_1 and R is now obtained using the root-locus phase rule, where θ is the angle from z_1 to R and where initially the phase contribution of the conjugate of z_1 is assumed to be $+90^o$. Thus,

$$-14^o + 45^o - 90^o - 135^o + \theta^o + 90^o = -180^o$$

$$\therefore \text{ hence,} \quad \theta = -76^o$$

Now knowing the approximate location of z_1 we are able to make a better estimate of the phase contribution of the conjugate of z_1 and on repeating the calculation we obtain a sufficiently accurate estimate of the angle as -75°. We now have the distance and angle of z_1 from R from which we compute the compensation zeros to be located at $-1.04 \pm j1.14$ in association with an additional gain of 24. The system must therefore be as shown in figure (5.7).

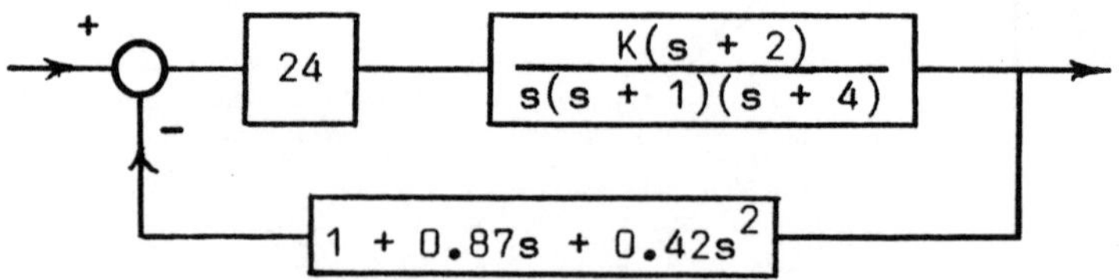

Fig (5.7): Completed design of the compensation to handle the variation shown in figure (5.6).

5.5. A REMINDER OF NOISE TRANSMISSION

It is quite possible to be carried away by the potential of feedback to solve sensitivity problems as shown in the examples of earlier sections and especially with the insertion of feedback zeros as with figure (5.7). However one's attention should not be distracted from the costs which have to be borne. These were mentioned in Section (4.4) and included difficulties which could be encountered when measurement feedback noise was in any way significant. Thus, repeating equation (4.11), the relationship between the plant drive signal θ_p and the feedback noise signal n is given by:

$$\frac{\theta_p}{n} = \frac{L(s)}{P(s)} \qquad 5.8$$

... over the higher frequency ranges.

That is, the transducer noise can undergo considerable amplification if the designer is prodigal with demands for sensitivity reduction with the likelihood of plant saturation and performance degradation. Further, if the numerator polynomial of equation (5.8) has an order equal to or greater than that of the denominator, then this amplification would theoretically exist up to infinite frequency with disastrous results should the noise even approximate to white.

If figure (5.7) is examined in this light it will be seen that just such a condition would exist. Note further that this will always be the case if the only compensation exists in the feedback path in the form of derivatives. Complete state-vector feedback solutions must therefore be viewed with some suspicion if the measurements are noisy.

It appears necessary therefore to include a forward-path compensation function in the design, not only to mould the closed-loop transfer function into the form which is required but also to introduce a sufficient number of poles within the loop to ensure that the feedback-noise transmission function falls to zero with increasing frequency. Alternatively, one might consider the construction of an approximation to the feedback function using available plant variables, with or without additional filtering. This could, at times, prove to be less costly and yet more effective.

5.6. S-PLANE DESIGN WITH FAR-OFF POLES

We have now seen that feedback zeros should be augmented with additional poles to minimise the noise problem with the consequent introduction of further complications in that the root-loci will become more complex and the system may lose its dependence on only a few dominant terms. That is, some of the introduced poles may, with the loop gain necessary to fix the dominant poles, migrate into the s-plane region occupied by those previously dominant terms. To ensure that this does not occur, it will be essential to locate them remotely so that they will not interfere with the dominance concept, but yet not so remote as to extend the loop bandwidth beyond what is necessary. This is yet another expression of the costs to be paid for the benefit of feedback.

Since the exact location of the additional poles is not critical, Horowitz (32, pp 249-264) suggests for ease of analysis that they should be multiple and so positioned that, with the loop gain required to achieve the dominant poles, they do not transgress some predefined dominance boundary which separates the dominant from the 'far-off' poles. Thus, using the example of section (5.4) on pages 91 and 92, we see that three additional poles are required (at least) and if an arbitrary dominance boundary is defined as the perpendicular through $s = -10$, say, the root-loci of these poles can be constructed approximately as shown in figure (5.8).

The triple pole is located approximately on the assumption that the three poles and zeros in the dominant region cancel as far as their effects on the root-loci are concerned. The loci will therefore be straight lines and for them to cross the dominance boundary at A and A', the maximum gain (2x24), being the product of the compensation gain of 24 and the maximum plant gain of 2, is given by:

$$K' = 48 = (pA)^3$$

The vector pA therefore has a length of 3.63 and since the locus makes an angle of 60° with the real axis, we have that p = −11.82. Of course, this is only approximate and the design can now be repeated including the the far-off poles in the calculation. Retaining however the present compensation, the loop can be arranged if required as shown in figure (5.9) which requires the availability of a rate signal only in addition to the system output. The closed-loop transfer function will have as a zero the pole of the transient rate term and a noise-transmission function which will decay towards zero at 6 db/octave.

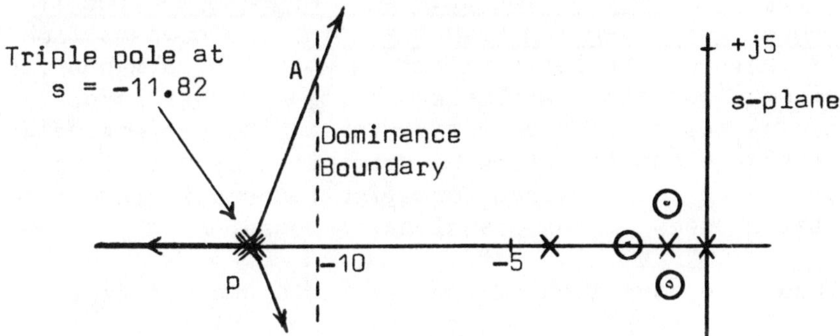

Fig (5.8): Approximate root-loci of triple pole to filter high-frequency measurement noise.

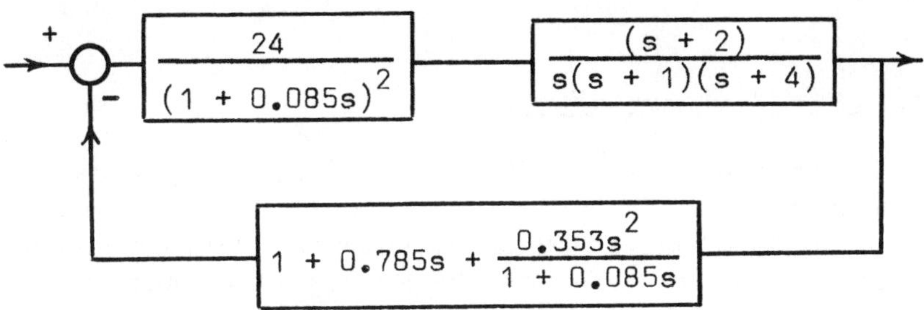

Fig (5.9): Feedback configuration obtained from s-plane design to achieve dominant-pole insensitivity incorporating rate and transient-rate compensation.

5.7. S-PLANE DESIGN WITH VARIABLE PLANT ZEROS

A problem arises when the zeros of the plant drift since these zeros will appear as the zeros of the closed-loop function and may, on drifting, result in quite large variation of the transient response. Thus, although the poles may have been made relatively insensitive by the feedback zeros, they determine essentially the modes of oscillation of the response, whereas the closed-loop zeros determine their magnitude.

The solution is simply to place a loop pole near to the offending zero thus forming a dipole which will have negligible effect on the response. Note that this is pseudo-cancellation since true cancellation would occur only when the zero had its nominal value. <u>When the zero drifts, the compensating pole will migrate towards it by an amount depending upon the gain</u>. The compensation will therefore be least effective when the gain has its minimum value and the dipole has its greatest separation. The two dipoles corresponding to the extreme values of the zero can therefore be determined by normal root-locus means and examined for effectiveness in reduction of the changes in the transient response.

Although we have treated the design in a piece-meal manner, considering first the dominant poles, then the far-off poles and finally drifting plant zeros, it must be approached with all these points in mind since later additions will obviously affect the earlier design decisions.

5.8. SUMMARY OF S-PLANE DESIGN

A design method has been presented which is based upon root-locus concepts and which is applicable to systems which can be modelled by relatively few dominant poles and zeros. As the order of the system increases within the region of dominance, so does the complication of the loci and the calculations which, since the method is basically one of cut-and-try, become rather cumbersome. More important however is the question of the ability of the designer to correlate pole-zero locations with the corresponding transient response since, as already pointed out, it is quite possible to vary the pole-zero array substantially and yet still achieve what is tantamount to the same response. We have also been reminded (Horowitz, 1963, pp 221-25, pp 266-7) that the demand for only a few dominant poles can be wasteful of plant capabilities and very expensive in gain-bandwidth requirements. However, when the plant is capable of relatively simple modelling, as a number are, and when the sensitivity specifications are not too demanding, the techniques outlined so far in this chapter provide the designer with a useful additional resource.

5.9. FREQUENCY RESPONSE SYNTHESIS

The second method for the synthesis of the loop function is based upon a consideration of the sensitivity of the closed-loop system function as a function of frequency and which is found to have definite advantages over the root-locus method.
The principal difference between the two lies in the specification of sensitivity which, for the frequency response approach, is given in terms of the allowable variation in the closed-loop frequency response over a given frequency range. The designer is thus able to control the variation in response of his system within what may be critical ranges and to permit wider variation where the frequency response has but little effect upon system behaviour.

5.10. GRAPHICAL SYNTHESIS OF THE LOOP FUNCTION

To over-come the practical limitations of the incremental sensitivity function, Horowitz (1963) introduced a large-scale sensitivity function $\tilde{S}_p^T(s)$ by considering the situation when the nominal plant $P_o(s)$ undergoes a large-scale change to $P(s)$, causing the nominal system function $T_o(s)$ to be modified to a new function $T(s)$.

$$\tilde{S}_p^T(s) \triangleq \frac{\Delta T(s)/T(s)}{\Delta P(s)/P(s)} = \frac{1}{1 + L_o(s)} \qquad 5.9$$

where $\Delta T = T - T_o$ and $\Delta P = P - P_o$

Manipulating the RHS of equation (5.9) leads to the fundamental relationship of equation (5.10) which provides the basis for the design procedure to be described.

$$\boxed{\frac{T_o}{T}(s) = \frac{\frac{P_o}{P}(s) + L_o(s)}{1 + L_o(s)}} \qquad 5.10$$

It would appear on examining equation (5.10) that the computational problem has, if anything, been considerably complicated. However, an examination of figure (5.10) illustrates the vectorial relationships of the above expression and which opens up the possibility of a graphical solution to the sensitivity problem.

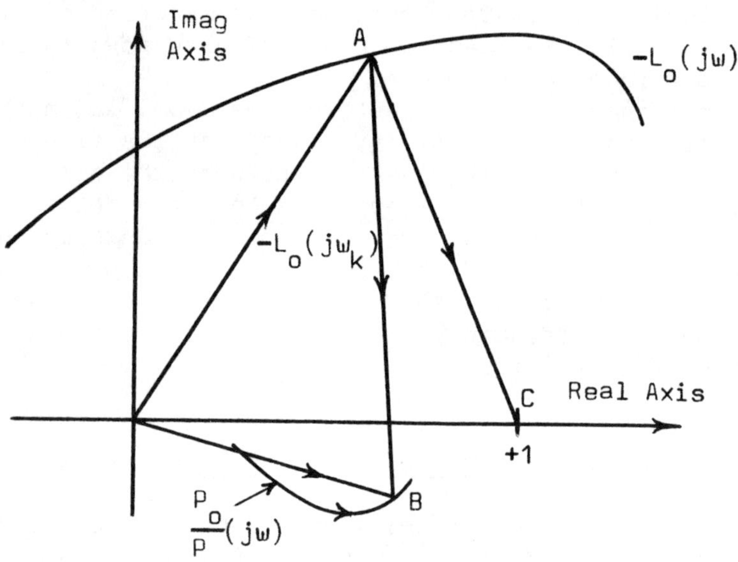

Fig (5.10): Graphical interpretation of the fundamental equation (5.10).

It will be seen that the following identities hold:

$$\underline{AC} = 1 + L_o(j\omega_k) \; : \; \underline{AB} = L_o(j\omega_k) + \frac{P_o}{P}(j\omega_k)$$

It is therefore obvious that:

$$\frac{T_o}{T}(jw_k) = \underline{AB}/\underline{AC} \qquad 5.11$$

It is therefore possible, given the system loop function on the assumption of nominal values, to compute the variation of the closed-loop function in magnitude and phase as a function of frequency for a given set of perturbed plant parameters. Alternatively, and of direct application to system design, if the vectorial relationships P_o/P and T_o/T are specified over a range of frequencies then the range of acceptable loop functions satisfying these relationships can be obtained graphically. This is illustrated in figure (5.11) which is drawn for one particular frequency w_k at which it is required that $a_1 \leqslant |T_o/T| \leqslant a_2$.

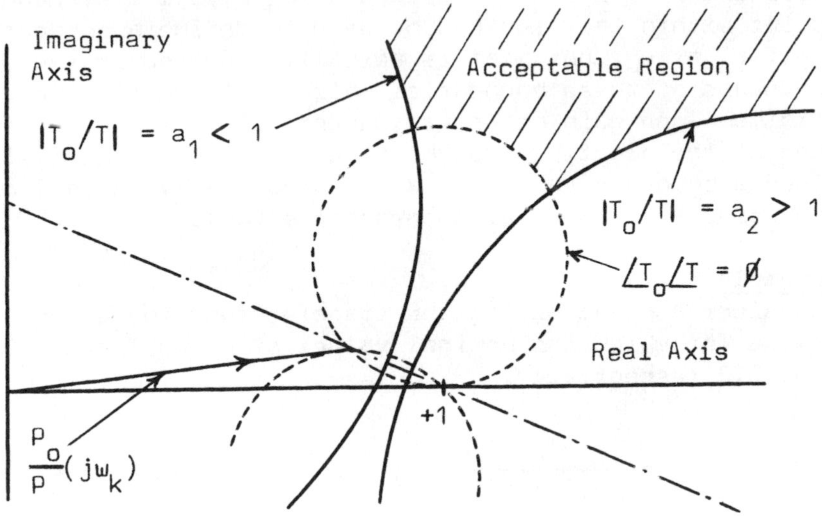

Fig (5.11): Definition of the acceptable region in which $-L_o(jw_k)$ must lie to ensure $a_1 \leqslant |T_o/T| \leqslant a_2$ and $\angle T_o \angle T \leqslant \emptyset$ at w_k.

The region lying outside the circles drawn using equation (5.11) corresponding to the magnitude ratio limits a_1 and a_2 is that which $-L_o(j\omega_k)$ must occupy if the closed-loop sensitivity specifications are to be satisfied. If there are also constraints on the permissible variation of the closed-loop phase values at ω_k then a similar set of circular boundaries can be drawn as shown in figure (5.11). (See Appendices for the definition of both sets of circles). Of course, the further out into the acceptable region the point corresponding to $-L_o$ at ω_k is the tighter will be the closed-loop variation - equivalent then to overdesign. The ultimate objective is of course to have $-L_o$ lying on the boundary at every frequency.

It must be noted that the magnitude and phase boundaries of figure (5.11) are applicable for one frequency only and for one particular plant ratio P_o/P. If, as is usual, the parameters of P(s) can vary randomly within their individual tolerance ranges then, rather than the point $P_o/P(j\omega_k)$ of the figure, there will be defined an area within which each variant of P_o/P will lie. Each point within the area can be used to define a boundary for $-L_o$ at ω_k, but what is actually required is the envelope of these boundaries which will specify the minimum loop gain necessary, under all possible variation of the plant, to desensitize the closed-loop system. Such a boundary can be obtained graphically (32,48) or by using a computer-aided procedure (49).

Example

Consider a plant having the transfer function given below for which the nominal values of K and T are 1.0 and 3.0 respectively.

$$P(s) = \frac{K}{s(1 + sT)} \; : \; \tfrac{1}{2} \leqslant K \leqslant 2 \; : \; 2 \leqslant T \leqslant 4$$

Let it be required to determine the constraints which must be imposed upon the loop function at the frequencies 0.1 and 0.5 rad/s (there would in practice be more specified frequencies) such that the tolerances on the magnitude of the closed-loop amplitude response are 10% and 20% respectively.

Before beginning the problem it is instructive to notice that although the closed-loop tolerances have been specified, the actual closed-loop transfer function has not been stipulated. This is a most useful facet of the technique and one which arises because of our intended use of a system with two degrees of freedom. That is, the loop function will be decided on the grounds of the sensitivity requirements (without reference to the system function) after which the distribution of the loop compensation around the loop is selected so as to arrive at a suitable closed-loop function. Of course, before we can define the tolerance range of the closed-loop frequency response which is required in the procedure, we must be aware to some degree of the form of system function which will be acceptable and the amount by which it can change and yet still have an operational system. More will be said on this point later.

To begin the example, the boundaries defining the range of the plant ratio P_o/P must be found by assuming a number of worst-case values for K and T and plotting the resulting polygon. Note that this will result in a pessimistic or conservative design since rarely will it be the case that randomly distributed parameters will reinforce one another in such a manner. For this very simple plant the polygons corresponding to the two specified frequencies are readily drawn as shown in figure (5.12).

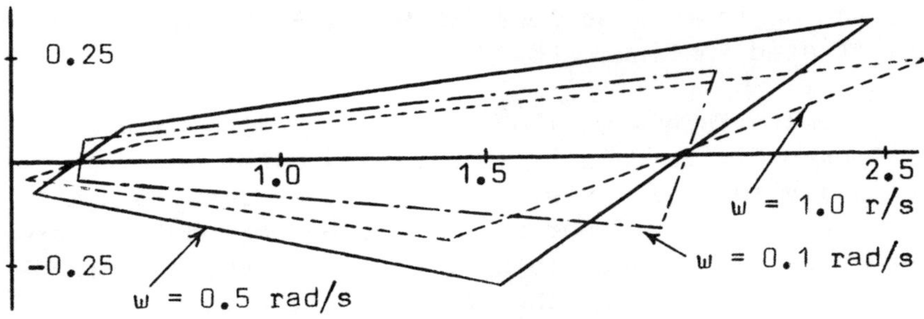

Fig (5.12): The approximate boundaries of P_o/P for the example on the basis of worst-case variations.

Next, it is required to determine the envelope of the circles, such as those shown in figure (5.11), which is obtained by drawing a few representative circles which have the largest radii corresponding to P_o/P points furthest from the point C (+1+j0). The vertices of the polygons of figure (5.12) are usually employed for this purpose. Two circles will be obtained for each point defined by the vectors $\underline{AB}$ and $\underline{AC}$ of figure (5.10) where:

$$\left|\frac{AB}{AC}\right| = \left|\frac{T_o(s)}{T(s)}\right| = \frac{1}{1 \pm m} \qquad 5.12$$

where m = fractional tolerance at ω_k.

The circles can be plotted by usual geometrical means or, more quickly and accurately, by calculating the centres and radii corresponding to the given m value. This procedure has been followed to produce figure (5.13a) which shows some of the circles for the present example for the frequency 0.1 rad/s.

From what has been said, a loop function which has sufficient magnitude (and suitable phasing) to lie outside all the circles, that is in the acceptable region, will result ultimately in a closed-loop frequency response with a tolerance of less than or equal to the specified ± 10% at 0.1 rad/s. We retain therefore only the envelope of the circles which now represents a loop magnitude boundary at this frequency. The above procedure is repeated for the frequency 0.5 rad/s with m = 0.2 and again the envelope is drawn. Both envelopes are illustrated in figure (5.13b).

The problem is now well on the way to being solved. All that needs to be done is to plot the plant function, as initially the only existing part of the loop, and to examine it in the light of the loop-magnitude boundaries to see what needs to be done in the way of compensation to ensure that the whole conforms to the constraints of the boundaries. For example, the loop function drawn in figure (5.13b) is acceptable since at the frequencies 0.1 and 0.5 rad/s the function lies without the boundaries. With more frequencies specified the loop function would of course be even more constrained thus limiting the options open to the designer.

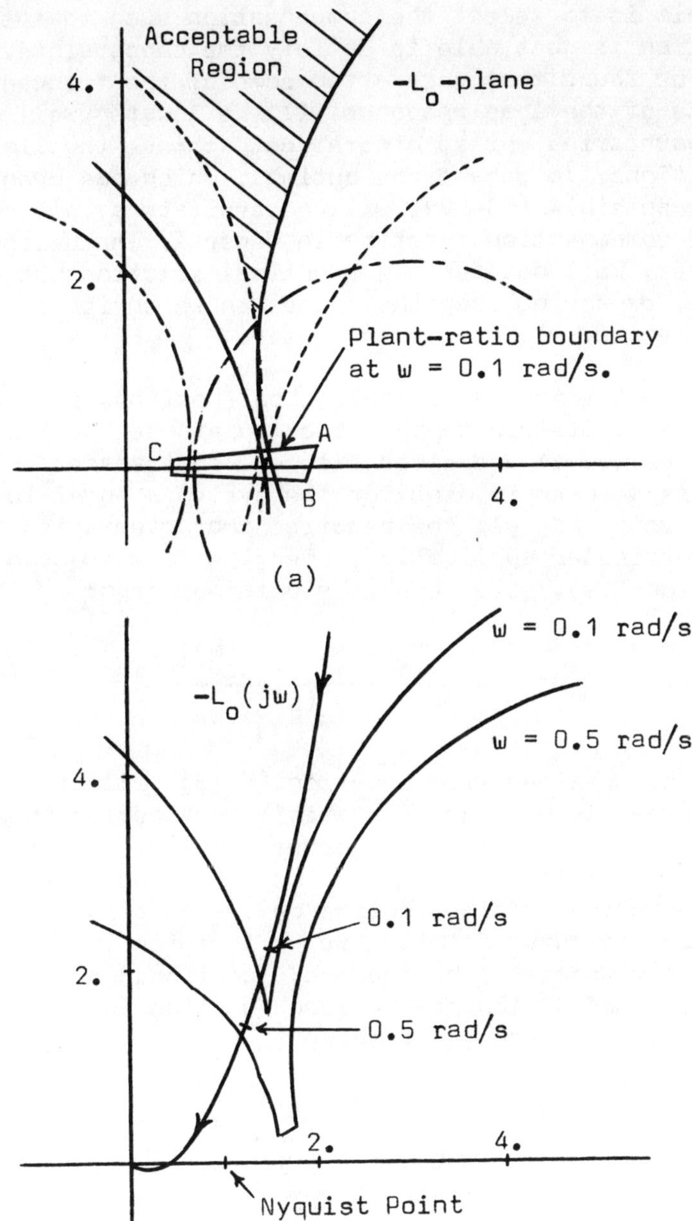

Fig (5.13): Graphical determination of the envelope of the loop magnitude boundaries of example. (a) the boundaries at 0.1 rad/s for the points A,B,C, and (b) the resulting boundaries and typical response.

The aim is to select the compensation such that the loop function is just able to satisfy the constraints. It will be found that very often some of the frequency points of the loop response will be located well within the boundaries whilst others just achieve the limiting conditions. To demand the optimal, which has been shown to be possible (50–52), will necessitate a very complicated compensation function in general. The designer may very well opt for a sub-optimal solution but one which, depending upon the manner in which it is to be implemented, might be less costly.

Having determined a suitable loop function, it now remains to distribute the function around the loop so as to arrive at a desired closed-loop transfer function. Suppose we have in mind for the system a model function $T_m(s)$ which has all the required characteristics for our particular application, then the forward-path compensation G(s), given L(s), is obtained from:

$$G(s) = \frac{T_m(s)\{1 + L_o(s)\}}{P_o(s)} \qquad 5.13$$

The G(s) obtained from equation (5.13) will often be of unnecessarily high order. Satisfactory performance will still be achieved if a low-order model of G(s) is used instead, obtained perhaps by employing the least-squares curve-fitting routine on the function G(jw), as long as the curve fit is good up to twice the bandwidth of $T_m(s)$. Finally, the feedback function H(s) is found from the definition of the loop function and the now known forward-path function.

$$L_o(s) = G_m(s).H(s).P_o(s) \qquad 5.14$$

where $G_m(s)$ is the low-order model of G(s).

H(s) may similarly be of high order and can also be modelled to provide a parsimonious function, but this time over the higher frequency ranges.

5.11. STABILITY AND THE VARIABLE PLANT

The previous section was concerned primarily with the determination of the loop function over a specified frequency range with no immediate thought given to the question of stability other than to ensure the the loop response approached the region of the Nyquist point within the first quadrant. However, with plants which involve variable parameters, it is not just a case of ensuring a satisfactory phase and gain margin but one of ensuring these criteria given the variations which can occur in the loop. Thus, we have:

$$L(s) = G(s)H(s)P(s) = L_o(s).\frac{P(s)}{P_o(s)} \qquad 5.15$$

We see that one particular realization of the loop function consists of the product of the nominal loop function and a vector which can have a considerable magnitude but which often has a limited phase variation at the higher frequencies we are considering since the phases of the nominal plant and any variant approach asymptotically the same phase value. A particular variant of the loop function may therefore encircle the Nyquist point causing instability even though the nominal loop function indicates a satisfactory system. Figure (5.14) illustrates an example of such a situation which obtained in the design of a ship autopilot (53) where the loop function had to be reshaped and its bandwidth extended to prevent any possibility of instability. Notice how the variational regions become more line-like as the frequency increases. By including a phase-advance network in the loop, the frequency response was modified as shown such that the worst-case phase margin was held at a suitable value.

5.12. SPECIFICATION OF CLOSED-LOOP TOLERANCES

The method of design adopted in this chapter has been one employing frequency-domain concepts and therefore requiring the demanded closed-loop sensitivity specifications to be defined in these terms. Bounds on the

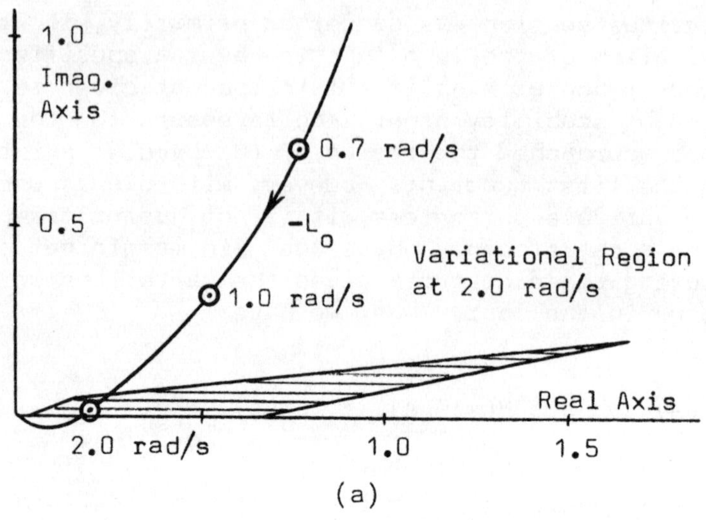

(a)

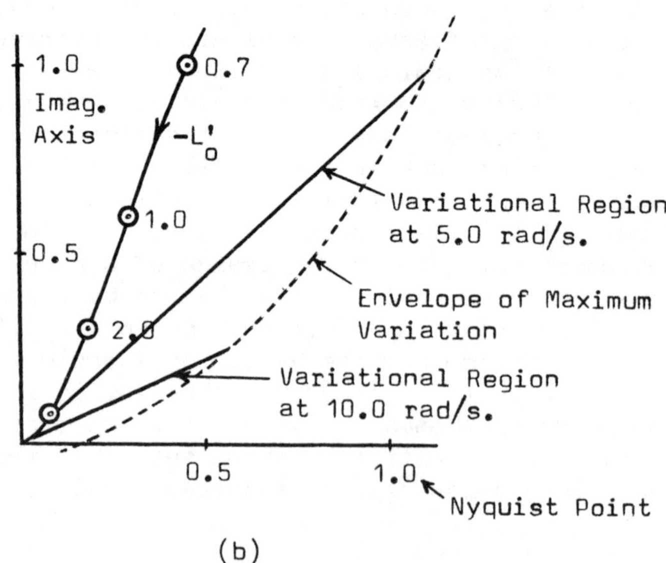

(b)

Fig (5.14): High-frequency variation of the loop function for a Mariner-Class vessel plus autopilot. (a) Indication of likelihood of instability due to plant variation. (b) Improvement, with extension of bandwidth, on introducing phase-advance term.

amplitude ratio of the frequency response are just as good as bounds on the transient response since, for minimum-phase systems, the definition of one uniquely determines the other. It may be however that prior operational performance has been decided in the time-domain which will require the translation of time-domain response boundaries into the form required for the application of the method of the previous sections. At present, the rigorous translation of such bounds has not been achieved.

Sidi (54) approaches the problem by assuming a dominant second-order model which adequately describes the proposed nominal time response of the closed-loop system and, from an examination of the variation of the transient response caused by variation of the parameters of the model, is then in a position to specify the tolerable variation of the frequency response.

Krishnan and Cruikshanks (55) endeavour to eliminate the trial-and-error necessary in Sidi's method by the exploitation of the condition:

$$\int_0^t \left\{ c(p,a) - m(p) \right\}^2 dp \leqslant \int_0^t v^2(p)\,dp \quad : \quad t \geqslant 0 \qquad 5.16$$

where p is a dummy time variable, $c(t,a)$ is the system response to an input $r(t)$ which is perturbed by plant variation 'a' and where $m(t)$ and $v(t)$ are defined in figure (5.15).

By assuming either that $m(t)$ is the nominal response or by incorporating a magnitude constraint on the sensitivity function to dampen disturbance response (48), the authors show that the condition of equation (5.16) reduces to a simple condition on the sensitivity function in the frequency domain. The difficulty with this approach is the need to know $v(j\omega)$ and $m(j\omega)$, a problem which is overcome only by resorting to modelling which the method was designed to replace, or on the assumption of analytic expressions for each of $b(t)$, $d(t)$ and $m(t)$. The latter may be satisfactory in some situations but can become unduly restrictive.

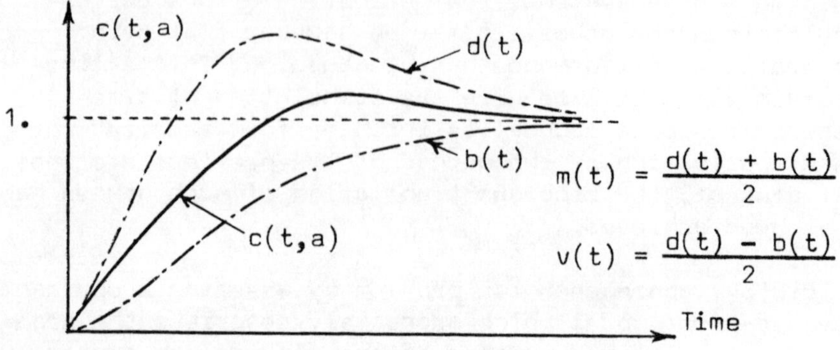

Fig (5.15): Time domain bounds on the system response as used by Krishnan and Cruikshanks (55).

The Sidi approach has been greatly extended (56) by using well documented models of standard form of up to third order and in particular by the use of the coefficient plane model (7,28). Rather than selecting a model because it has a transient response which can be accepted, the model can be determined by an optimization routine (9) which carries out the selection on the basis of the best fit to a number of dynamic specifications, since these are decided by the system function and not its structure. Having determined the model, and knowing the time-domain boundaries, the allowable variation of the model parameters is found by use of the transient response sensitivity function of equation (4.21) and the tolerance expression of equation (4.25). Knowledge of the model parameters and their variances allows the use of a Monte-Carlo determination of the allowable spread of the frequency response.

5.13. LOOP SYNTHESIS VIA THE NICHOLS CHART

An alternative approach to loop-function synthesis, and one which is somewhat less laborious when carried out by hand, is to perform the design using a Nichols chart where all magnitudes, open and closed-loop, are measured in logarithmic terms. Since the curves of this chart have been plotted on the assumption of unity feedback, we must write the general form of T(s) as:

$$T(s) = \frac{G(s)P(s)}{1 + L(s)} = \frac{1}{H(s)} \cdot \frac{L(s)}{1 + L(s)} \qquad 5.17$$

Now, writing equation (5.17) in logarithmic form:

$$20\log_{10}T = 20\log_{10}\frac{L}{1 + L} - 20\log_{10}H \qquad 5.18$$

... from which we obtain:

$$\Delta(20\log_{10}T) = \Delta(20\log_{10}\frac{L}{1 + L}) \qquad 5.19$$

... and of course:

$$\Delta(20\log_{10}L) = \Delta(20\log_{10}P) \qquad 5.20$$

We see from equations (5.19,20) that the specified allowable variation of the closed-loop function $20\log_{10}T$ is directly related to the variation of the function $(L/(1 + L))$ and that changes in the magnitude of the plant follow exactly those changes in the loop function. These simple relationships readily enable the determination of the loop-magnitude boundaries with the aid of templates of the plant variation at each of the specified frequencies (48,54).

Thus, consider a plant having the range of variation ABCD shown in figure (5.16). This particular plant is in fact that used as an example in section (5.10) on page 100 at the frequency 0.1 rad/s, where the closed-loop tolerance of $\pm$ 10% has been converted for use on the Nichols chart to the equivalent range of variation of 1.75 db. Hence, from equation (5.19), the closed-loop magnitude range can be obtained by noting the range of the chart curves covered by the template of plant variation. Thus, for the uncompensated plant, this range is much less than 1.75 db indicating that the loop function is unnecessarily high. On shifting the template to any of the other locations illustrated

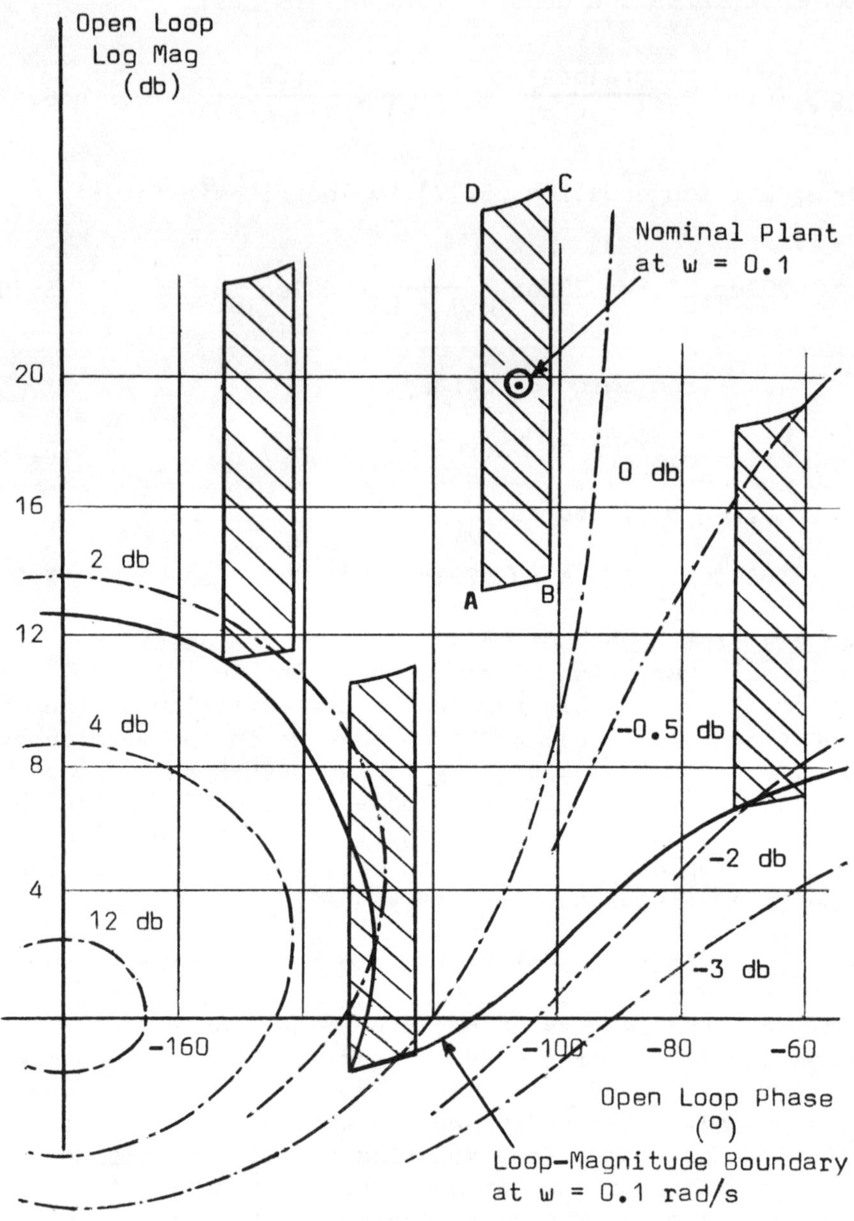

Fig (5.16): Determination of the loop-magnitude boundary at one specific frequency via the Nichols chart. No closed-loop phase curves are shown.

it will now be seen that the range of closed-loop magnitude variation takes up the maximum allowable of 1.75 db. A curve has been drawn in the figure which joins all those points corresponding to vertex A of the template for which the closed-loop variation of 1.75 db applies. This curve represents the minimum bounds of the loop function when the plant is in condition A. To obtain the minimum bounds of the loop function in its nominal state, the plotted boundary curve must be translated in magnitude and phase by amounts equal to the translation of point A to the nominal point of the plant template.

The design procedure is thus quite straightforward, providing the designer with the insight to make whatever trade-offs are required and requiring only the tracing of the plant variation templates and their manipulation on a Nichols chart.

5.14. COMPUTER-AIDED DESIGN

The computational problem posed in frequency-domain synthesis, involving as it does the statistical variation of a number of plant parameters, is one which is eminently suited to solution by computer using quite straightforward and simple algorithms. The flow diagram of one possible procedure is shown in figure (5.17) in which the operator interacts in a conversational manner in response to visual displays of the loop function and its sensitivity boundaries (49). The operation can be divided into three stages.

5.14.1. Examination of plant variation

Initially the plant variation must be examined to ascertain if further attention need be given to sensitivity considerations and, if it is, to determine at every frequency over which the specifications are given the maximum likely range of that variation. It is considered that the simplest approach to achieve this objective is to select, say, 1000 plant variants in a Monte-Carlo manner, that is, by perturbing the plant parameters by adding to the nominal value of each parameter a factor

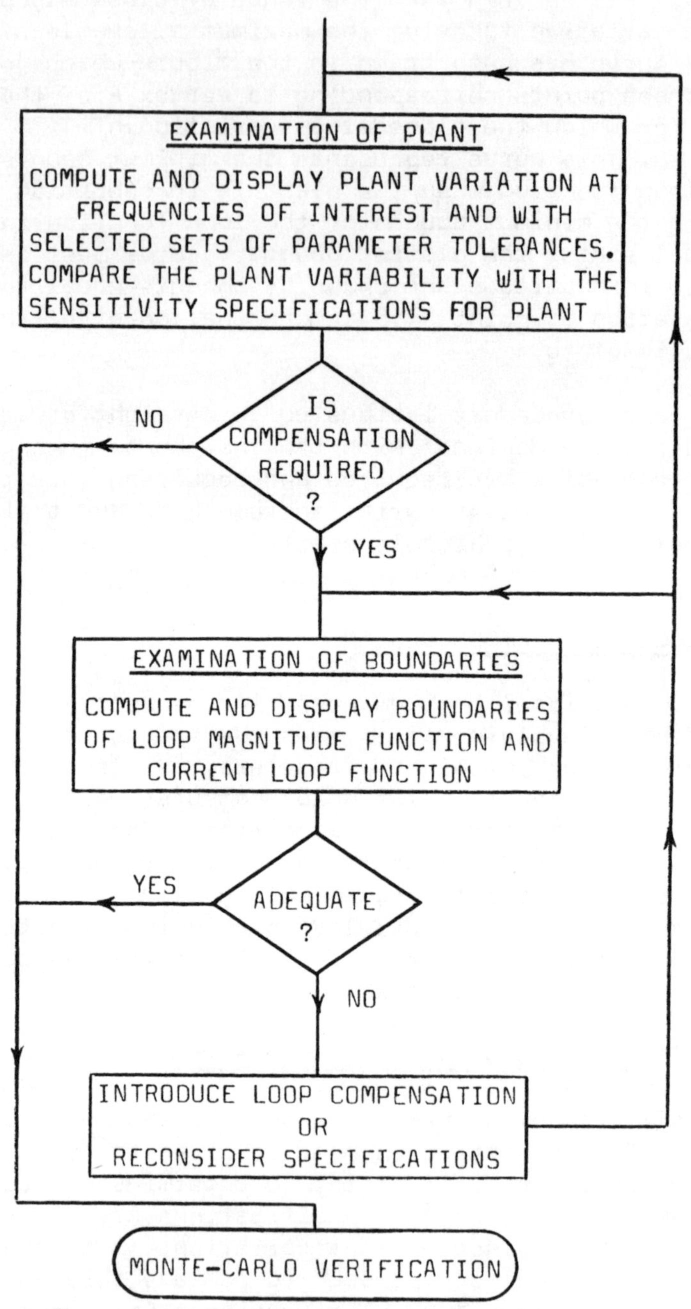

Fig (5.17): Flow diagram of computer-aided synthesis procedure for the loop function.

composed of the product of the tolerance imposed upon that parameter and a random number. The number may be drawn from a uniform distribution (which will give a greater spread and ultimately a more conservative design) or, as is more common, from a Gaussian population obtained by processing uniformly distributed numbers. Methods of random number generation with given probability distributions are covered in standard texts (44). Rather than proceed with the computation of the boundaries using all 1000 values of the ratio $P_o/P(j\omega_k)$ at each frequency, it is more economical to determine a polygonal boundary, such as that of figure (5.18a). With the routine of reference (49), a polygon having twenty vertices is constructed about the point $P_o/P = 1$, each of which represents the maximum ratio within the twenty equal sectors about that point. With each of of these values as a vertex, an area is defined in the P_o/P plane within which lies, to an adequate approximation, all possible plant variants at that particular frequency.

5.14.2. COMPUTATION OF THE BOUNDARIES

Referring to figure (5.11) it is seen that it will be necessary to compute the circular boundaries which correspond to the closed-loop specifications for each of the twenty representative plants for each of the stipulated frequencies. Having done this, it is then necessary to obtain the envelope at each frequency which encloses the union of the circles. It is therefore a simple matter to obtain this envelope in terms of the real-axis co-ordinate by retaining only the maximum ordinates of the twenty circles, with the real-axis values chosen at the discretion of the designer.

5.14.3. DETERMINATION OF THE FEEDBACK

On first using the programme, the loop-magnitude boundaries will be displayed together with the function $-P_o(j\omega)$ which represents the loop function prior to the introduction of compensation. Should the point $-P_o(j\omega_k)$ lie beyond the loop-magnitude boundary at ω_k and should this be so for all frequencies, then no further action is required. However, the usual situation is that some of the points will lie within their associated envelopes and it will be necessary to involve some compensation terms to mould the shape of the loop function to correct for the infringement of the constraints. After each modification of the loop, it will be displayed with all the boundaries. This procedure can be repeated until the designer is completely satisfied that he has the 'best' loop function. The programme ends with the computation of the limits of the closed-loop frequency response by Monte-Carlo means, again say for 1000 possible system functions, in order to validate the design.

Experience has shown that even a minimal quantity of software can result in significant time-saving and yield dividends far in excess of that expected by such a limited committment to programme development. However, there is always the danger of over-committment to convenient algorithms which, while relieving the user of trivial but laborious tasks, might also tend to relieve him of the need to use his judgement whilst simultaneously, to ease the burden of programming, imposing what may be unjustified constraints on the design. This would indeed be a misfortune with a synthesis procedure of such transparency. It is better, bearing in mind the advice of Bode, to leave judgements and the final synthesis in the hands of the designer while employing the rules (algorithms) to lay stress on the concepts and processes.

5.15. OPTIMIZATION OF THE LOOP FUNCTION

It is obvious from an examination of a typical set of loop-magnitude boundaries that a multitude of loop functions can be devised which will satisfy the sensitivity specifications. It has also been observed that a fundamental dilemma in feedback system design arises from the opposing demands made by sensitivity and measurement noise reduction on the extent of the loop bandwidth. Loop functions which lay on the boundaries at their corresponding frequencies would be desirable but, for any arbitrary set of bounds, can such a rational function be defined? Additionally, on the assumption

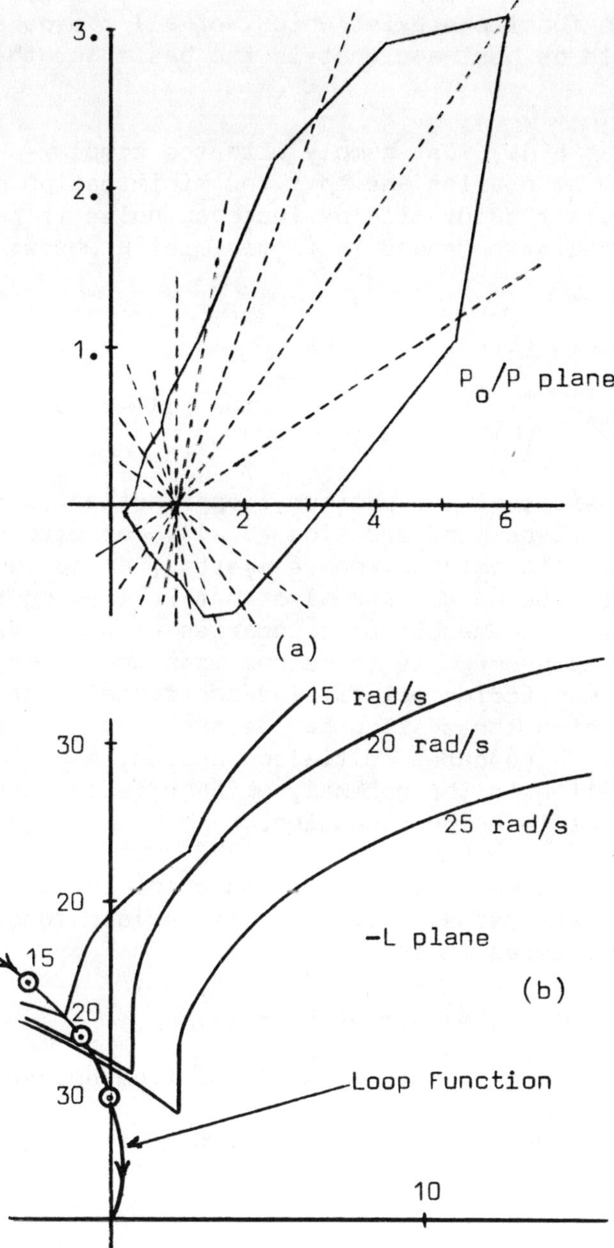

Fig (5.18): Typical P_o/P display (a) and loop-function display (b), together with loop-magnitude boundaries, generated by computer-aided design programmes.

that such functions exist which, of all the possibilities, would be best and what is the basis for the selection?

Given that L(jw) must comply with the complex-plane magnitude boundaries and that the minimisation of the risk of plant saturation by feedback noise is paramount then a realistic demand is to minimise K where:

$$\lim_{s \to \infty} \{L(s)\} = Ks^{-m} \qquad 5.21$$

where m is the pole-zero excess of L(s).

On the basis that the optimum loop function is that which minimises K of equation (5.21), Horowitz (50) shows that, for minimum-phase systems, an optimum does exist which is unique and also that it lies on the associated loop-magnitude boundaries at every frequency. Further improvement is to be had when the plant has internal monitoring points. The additional degrees-of-freedom which these allow can be employed (57) to synthesize a cascaded multi-loop design, each loop approximating to the optimal, which effects further improvement in noise reduction.

Further problems occur should the plant include non-minimum phase terms. Thus, consider a loop function having two zeros in the RHP:

$$L(s) = L_1(s)(1 - sa)(1 - sb) \qquad 5.22$$

where $L_1(s)$ is minimum phase.

The expression for L(s) may be written in an alternative form:

$$L(s) = L_1(s)(1 + sa)(1 + sb).\frac{(1 - sa)(1 - sb)}{(1 + sa)(1 + sb)}$$

$$= L_m(s)A(s) \qquad 5.23$$

A(s) is an all-pass function so that whilst L(s) and $L_m(s)$ have the same magnitudes, they differ in phase by that of A(s). Now the latter has a phase which always decreases with frequency so that the increase in phase lag of L(jw) is greater that that of $L_m(jw)$ by an amount argA(jw). Thus, $L_m(s)$ is said to be a minimum-phase function since, for its given magnitude characteristic, it has the least phase, whilst obviously L(s) is non-minimum phase. In a similar manner, a loop function containing a pure delay element can be thought of as non-minimum phase (NMP) since it can be closely approximated by an all-pass function with a phase lag which is proportional to frequency.

The problem which such elements introduce is that for given stability margins there is a limited amount of phase lag allowable from which must be subtracted that additional lag of A(s). Thus, to keep within bounds, the rate of decay of $L_m(s)$ must be reduced below what might normally be expected which in effect implies that it must be controlled over a greater frequency range. Nevertheless, Horowitz and Sidi (51) were able to extend their synthesis procedures and show that if an optimum existed for a NMP plant then it would be unique and would lie, at each frequency, on the corresponding boundaries. Their proof required, however, the introduction of two restrictions:

(a) Any RHP plant pole must be specified with a region of uncertainty which included areas of both the LHP and the RHP.

(b) The compensation terms of the loop function must be minimum-phase and stable.

It is possible however that for any particular NMP plant there will be no optimum solution. The authors present criteria to determine if this is the case.

While the results of the research outlined above is interesting, it is not immediately applicable to design problems. A recent contribution (52) by Gera however presents a numerical technique which is reputed to be fast to enable the derivation of an approximation to the optimum loop function. Although the information is in the form of frequency-response data points it does give an indication of the best which can be achieved. As can be imagined, given an arbitrary plant and sensitivity specifications, a rational function approximation to the optimum could be quite complex and has not yet been solved. However it should be possible to deduce satisfactory guide-lines by the simple expedient of least-squares complex-curve fitting (20-22).

5.16. APPLICATIONS TO NONLINEAR SYSTEMS

The tacit assumption has been made in the earlier material that each plant has been linear, a convenient fiction perpetuated to simplify the mathematical manipulations and in order, often, to reach an analytic solution. There are many situations in which the assumption of linearity can be justified on the grounds that the deviation of the actual response from the modelled response is negligible for sufficiently small signal levels. Whether the nonlinearity will have a significant effect on the performance of the system may well depend upon the demands made or the actual mode of operation.

It has been observed that the distortions caused by a nonlinearity can be reduced by the use of feedback, that is, that it has a linearising effect. One can think of some nonlinearities in terms of a linear element with parameters which exhibit sporadic variation. Viewed in this light, the problem of nonlinear compensation appears as one of sensitivity reduction which, as has been discussed, can be solved by invoking a feedback configuration. This is not to suggest that all problems of nonlinear behaviour can, or should, be handled in this manner.

5.16.1. An Example of Uncertain Nonlinear Dynamics

The equations of motion of a ship in a horizontal plane are readily expressed (58) in terms of the hydrodynamic surge force X, sway force Y and the yawing moment N, which are complex nonlinear function of the ships motion.

Abkowitz assumes they are nonlinear functions of the rates and accelerations of the variables of motion and proceeds to expand them in a Taylor's series about a steady-state condition, usually one of constant forward speed in a straight line. By terminating the expansions of these equations (5.24) after the third term he arrives at a nonlinear model which adequately predicts complex manoeuvring behaviour.

$$\left.\begin{matrix} X \\ Y \\ N \end{matrix}\right\} = f(u,v,r,\dot{u},\dot{v},\dot{r},\delta) \qquad 5.24$$

where u,v,r,δ represent forward velocity, sway velocity, yaw rate and applied rudder angle.

The coefficients of the terms of the expansions can be determined experimentally by carrying out captive-model tests (59) and measuring the forces and moments applied to the model hull to engender certain evolutions. Once determined, the equations of motion can be manipulated into the form of a transfer function relating, say, the ship yaw to the applied rudder angle which, although of linear form, has parameters which are nonlinear functions of ship motion.

One way to approach this problem is to carry out some simulations of the ship motion, using the full set of Abkowitz's expansions for a whole range of operational demands as might be expected when manoeuvring and hence to compute the variation of the parameters of the equivalent transfer function as functions of time. Thus with the constant-speed straight-line motion parameters as nominal, a probabilistic transfer function can be defined which is assumed linear but whose parameters are uncertain (in that they are manoeuvre-dependent) and can lie anywhere within the ranges determined from the simulations (60). Thus, for the Mariner-Class vessel, the model listed in Table (5.1) is found to apply. Using this model and the loop-synthesis methods discussed, the design of an autopilot for the ship has been reported elsewhere (53). A typical result of a course-change manoeuvre is however shown in figure 5.19

TABLE (5.1): Definition of the probabilistic model representative of the Mariner-Class of vessel

$$P(s) = \frac{K(1 + s\tau)}{(1 + \frac{2\zeta s}{\omega_n} + \frac{s^2}{\omega_n^2})}$$

Parameter	Nominal Value	% Tolerance Range	
K	−0.1837	+6.2	−93.2
τ	+18.530	+1.7	−32.0
ζ	+2.0707	+15.5	−52.6
ω_n	+0.0331	+317.2	−17.2

in comparison with that obtained when the ship was under the control of a self-tuning controller (61). It was assumed for the purposes of design that the variation of the parameters of Table (5.1) was random and that each variate was uncorrelated. This is not strictly correct as can be expected when the transfer function parameters are themselves constructed from a set of hydrodynamic derivatives. Indeed, relationships can be derived between these parameters which, if used, would limit the total range of variation and hence the loop bandwidth requirement. Nevertheless, the reduction which has been achieved in the variation of the course-change response is quite satisfactory.

5.16.2. The use of Nonlinear Controllers

The method of Section (5.16.1) may result in serious feedback noise problems should the allowance for nonlinear behaviour require a large range of variation of the equivalent linear plant. One possible means of avoiding this situation, and one where the method is feasible given the increased availability of digital

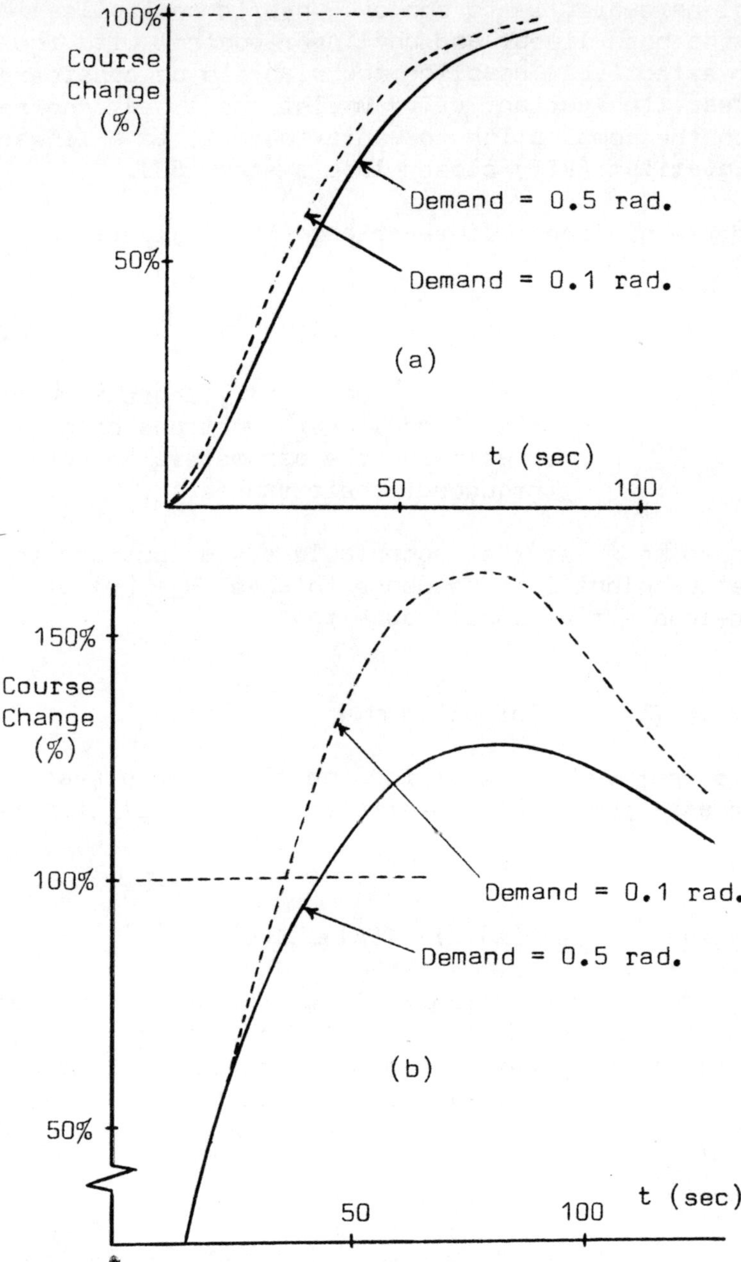

Fig (5.19): Response to corse-change demands for (a) a design based upon system sensitivity reduction and (b) a self-tuning controller.

control hardware, is to employ a hybrid controller involving both linear and nonlinear control with the latter effectively enabling the plant to be considered as linear time-variant (LTV) whilst the linear control acts on the combination to approximate it to a linear time-invariant (LTI) closed-loop system (62).

Consider a nonlinear time-variant (NLTV) plant:

$$y = q(x) \quad : \quad N(y) = M(x) \qquad 5.25$$

where $N \in \mathcal{N}$ and $M \in \mathcal{M}$, where $\mathcal{N}, \mathcal{M}$ are sets of nonlinear functions obtained by allowing the parameters to vary throughout their ranges.

Let there be a set $\mathcal{Y}$ of acceptable system outputs for the set of plant $\mathcal{Q}$ in response to a set $\mathcal{R} = \{r\}$ of closed-loop system inputs such that:

$$y \in \mathcal{Y} \quad : \quad \text{for all } q \in \mathcal{Q} \qquad 5.26$$

This is achieved by requiring the closed-loop system to behave as a member of a desirable set $\mathfrak{T}$ of LTI functions:

$$y(s) \triangleq T(s)r(s) \quad : \quad T(s) \in \mathfrak{T}(s) \qquad 5.27$$

The set $\mathfrak{T}$ is found by conventional methods by suitable tolerancing of the parameters of T(s). Now to convert from NLTV to LTI, define a nominal output y_o by the selection of the nominal N and M such that:

$$y_o = h * u = v \quad : \quad q = q_o \qquad 5.28$$

where $h = \mathcal{L}^{-1}H(s)$ and H(s) is the desired LTI plant function after nonlinear compensation.

We now construct a cancellation function:

$$x = \Lambda(v) \quad : \quad N_0(v) = M_0(v) \qquad 5.29$$

In essence, a model-following system has been defined with y(t) following v(t) exactly whenever the plant assumes its nominal form, with the nonlinearities of q being compensated by those of Λ. The modified plant is therefore:

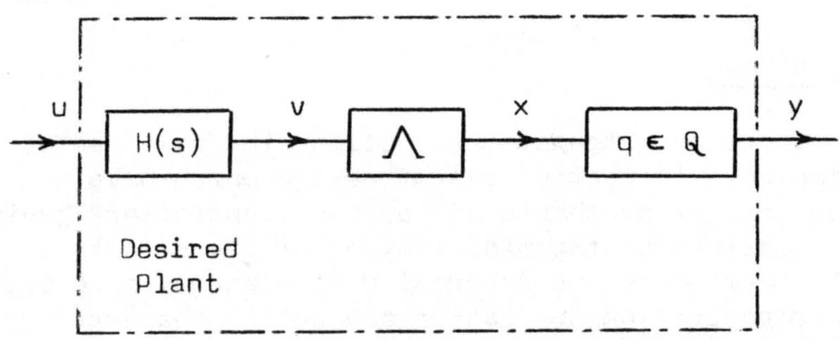

Modified Plant

Fig (5.20): Nonlinear cancellation compensation to limit the variability of the modified plant prior to LTI design.

The modified plant of figure (5.20) therefore shows variations from H(s) only if the parameters of the NLTV plant change, not because of the nonlinearities themselves. We now have:

$$\frac{y}{u}(s) = P(s) \quad : \quad P \in \mathcal{P} \qquad 5.30$$

The extent of the set $\mathcal{P}$ must now be found before proceeding on the basis of a LTI sensitivity design. To do this, the known members of the set $\{y\}$ are substituted into equation (5.31) which is solved numerically for v(t) and v(jw).

$$N(y) = M(N_0 v) \quad : \quad \mathcal{P} = \left\{\frac{y(jw)}{v(jw)}\right\}.H(jw) \qquad 5.31$$

The advantage of this approach is that, for a large class of nonlinear systems, the cost of feedback can be greatly reduced since, in the absence of uncertainty, there is no need for the provision of feedback to handle the vagaries of nonlinear behaviour. That is, the modified plant behaves as if it were LTI for the stipulated set of system inputs. The feedback is required only to limit the effects of variation of the coefficients of N and M (63).

5.17. SUMMARY

The chapter has attempted to outline the most useful developments in control system design which have influenced the synthesis of feedback controllers having the objectives of the minimization of the effects of plant uncertainty and external disturbances (more will be said concerning the latter subject in the next chapter) on system behaviour. It is believed that these are the most important problem areas, the difficulty of which increases rapidly as the complexity of the plant increases.

Two design methods have been presented. The first, based upon root-locus concepts, finds its main application when the plant is relatively simple and has been summarized on page 96. The second, based upon frequency-domain concepts, is widely applicable and conceptually simpler than the former method. With little effort, it is possible to speed-up the graphical manipulations which it entails by performing all operations on a computer with VDU facilities.

Emphasis has been placed throughout on the single-loop system in the interests of clarity and, while the techniques are directly applicable to multi-variable systems designed initially on the basis of non-interaction or diagonal dominance (64), interest has been shown in this area for a considerable time (32). More recently, functional analysis has been called upon to extend the design capability (65) of the method to multi-variable systems which, besides being uncertain, are also time-varying and nonlinear. Mention must also be made of the work of Sidi (36) who has adapted loop-synthesis to the particular problems to be met with in the design of sampled-data control systems.

Although the application of functional analysis is an exciting development with the promise of solution of the most difficult problems, it appears to have a serious shortcoming in that large LTI sets of P(s) are generated which could result in uneconomic designs. Research is continuing in this area.

In conclusion, it is the authors opinion that there is as yet no viable alternative to the use of the frequency domain for sensitivity-reduction design. State-variable formulation has little of significance to contribute in this area, obsessed as it is with differential eigenvalue and eigenvector sensitivity, usually quite unrelated to system response tolerances and to the problem of large plant parameter uncertainty (57). What results have been achieved are imbedded in such complex expressions as to find little application in synthesis (38). This contrasts adversly with the familiarity, simplicity and transparency of frequency domain concepts.

CHAPTER 6

Noise Disturbance

6.1. INTRODUCTION

The term 'noise' is employed universally to refer to those spurious signals which often appear as the higher frequency contamination seen superimposed upon the useful signals of the loop and which are instrumental in the degradation of the attainable accuracy and system performance. As with parametric variation, noise is ever-present, arising internally and externally from numerous sources including pick-up interference, inherently noisy electronics and instrumentation, external influences (eg, the radar glint effect) and the quantization noise of digital devices.

Obviously every care should be taken by the designer to eliminate such sources where possible, or at least to minimise their influences by screening or perhaps by re-siting. It may also be necessary to use higher grade electronics and transducers if the performance demands are stringent. Certainly he should not exacerbate the problem as exemplified by an in-service servomechanism of which the author is aware where dither was incorporated to mitigate against stiction only to give rise to gross levels of output noise. The cause of the dither amplification was the existence of a secondary resonance within the hydraulic drive with a resonant frequency in the neighbourhood of the dither frequency. In some cases the noise levels may be reduced by filtering but if the noise is wide-band this will cause distorted following (on filtering the input signal) or result in unavoidable constraints on the loop function. This in turn will limit the benefits achievable in the area of sensitivity reduction.

Noise has three primary effects on the operation of the system. Firstly, it will be transmitted to the output and thereby cause it to deviate from its demanded value, adding to system inaccuracies. Secondly, the noise will tend to increase power consumption and mechanical wear-and-tear as the plant endeavours to follow the rapid fluctuations. Thirdly, if the noise levels are very high, the plant may be forced to operate repeatedly under saturated conditions resulting in deterioration of performance.

The remainder of this chapter will be devoted to a brief discussion of a number of techniques which designers have found useful in their dealings with the noise problem. The techniques will not pave the way for the total elimination of noise but will enable the designer to effect some reduction in its level and to be aware of the pitfalls and trade-offs open to him.

6.2. ESTIMATION OF NOISE LEVELS

When noise enters the system with the command signal we are particularly interested in the noise levels to be expected at the output, to infer the additional inaccuracy, and at the plant input, to estimate the likelihood of saturation and power expenditure. The relevant transfer functions are:

$$\frac{\theta_o}{\theta_i}(s) = T(s) \quad \text{and} \quad \frac{\theta_p}{\theta_i}(s) = \frac{T}{P}(s) \qquad 6.1$$

where θ_i, θ_o and θ_p are the input, output and plant-input signals.

Here we consider θ_i to consist solely of the noise signal which, for mathematical tractability, we will assume to be stochastic and stationary, derived by passing white noise through a linear filter. The concept of white noise, although impractical, is very useful and implies that all frequencies are present in the signal which has a constant power-density spectrum. Experimentally, the power contained in a narrow band of frequencies is determined by applying the signal to a frequency-selective filter followed by a power measuring device. As the tuned frequency of the filter is changed the output power will vary. The variation of power with frequency is a measure of the power spectrum of the signal.

(Note that although it is not feasible to generate white noise, since infinite power would be involved, a given signal may be considered as white if its bandwidth is more extensive than that of the system to which it is applied).

The use of Fourier analysis is ruled out for the calculation of spectra since the condition which must be satisfied by the signal x(t) for the method to be applicable is:

$$\int_{-\infty}^{\infty} |x(t)|\, dt < \infty \qquad 6.2$$

Expression (6.2) is obviously not satisfied by any stationary stochastic process and it appears that Fourier techniques cannot be used to obtain spectral distributions for such processes. However, the autocorrelation function $R(\tau)$ which relates the current value of the variable to values at a future time τ, as a function of the original time series will reflect the frequency content of the signal (67). If x(t) has a zero mean value then $R(\tau)$ will approach zero as τ approaches infinity and equation (6.2) will be satisfied when x is replaced by R. That is, Fourier techniques have been re-instated. Beginning with the autocorrelation function, it is readily shown (67) that it is related to the power-density spectrum $\Phi(\omega)$ as a Fourier Transform pair (see equation 6.3). Notice that for $\tau = 0$, the corresponding value of the autocorrelation function R(0) is equal to the mean-square value of the

$$R(\tau) = \frac{1}{2\pi j}\int_{-j\infty}^{j\infty} \Phi(s)e^{s\tau}ds \qquad 6.3(a)$$

$$\Phi(s) = \int_{-\infty}^{\infty} R(\tau)e^{-s\tau}d\tau \qquad 6.3(b)$$

signal, which can be interpreted as the power of the signal and which is equal, using equation (6.3a), to the integral over all frequencies of $\Phi(\omega)$, thus confirming the definition of the power spectrum. That is:

$$\overline{x^2(t)} = R(0) = \frac{1}{2\pi j}\int_{-j\infty}^{j\infty} \Phi(s)ds \qquad 6.4$$

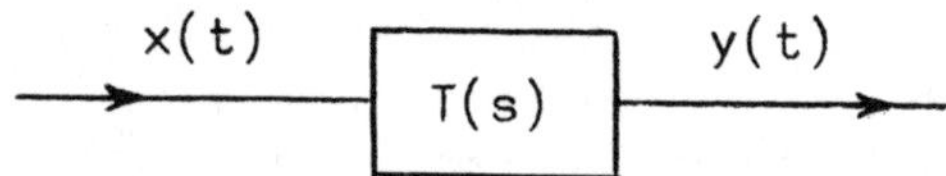

Fig (6.1): The linear system acting as a filter to random excitation x(t).

Before continuing to calculate the power or mean-square values of noise we must relate the spectra of the signal y(t) of figure (6.1) to that of the input x(t) to the filter (or system) T(s). Newland (67, pp 71-3) shows that the following relationship applies:

$$\Phi_y(s) = T(s)T(-s)\Phi_x(s) \qquad 6.5$$

As an example, suppose that examination of the input noise spectrum indicates that a suitable noise model consists of white noise of unity power density which has been filtered by a function 2/(1 + s) and that the system T(s) is given by 1/(1 + 2s). We therefore have:

$$\Phi_x(s) = \frac{2}{(1+s)} \cdot \frac{2}{(1-s)} = \frac{4}{(1-s^2)}$$

The mean-square value of the output y(t) is therefore:

$$\overline{y^2(t)} = \frac{1}{2\pi j} \int_{-j\infty}^{j\infty} \frac{4}{(1+s)(1-s)(1+2s)(1-2s)} ds$$

$$= \frac{1}{2\pi j} \int_{-j\infty}^{j\infty} \frac{2}{(1+s)(1+2s)} \cdot \frac{2}{(1-s)(1-2s)} ds$$

The solution of integrals of the above form have been tabulated (66) and are considered in Appendix (D). They are to be partitioned into the form of equation (6.6) when the integral is then given in terms of the coefficients c and d.

$$I_n = \frac{1}{2\pi j} \int_{-j\infty}^{j\infty} \frac{c(s)c(-s)}{d(s)d(-s)} ds \qquad 6.6$$

Thus, for our problem, we are given that:

$$I_2 = \frac{c_1^2 d_0 + c_0^2 d_2}{2 d_0 d_1 d_2} = \frac{2}{3} \text{ units}$$

where the coefficients are given by:

$c_0 = 2$, $c_1 = 0$, $d_0 = 1$, $d_1 = 3$, $d_2 = 2$

To see how well the filter has performed in attenuating the noise, we can find by the same means the mean-square value of the input noise. The white noise filter is of first order, hence:

$$I_1 = c_0^2/2d_0d_1 = 2 \text{ units}$$

The filter (or system in our application) has attenuated the signal noise to one third of its former power.

Of course we could equally well have calculated by the above method the power of noise at the plant input by using T/P in equation (6.1) instead of T. As seen above the method is quite straightforward but it does assume that the system is linear and that the noise model is stationary. Since the system transfer functions are taken as known, the application of the technique is one of analysis rather than design. It should be noted from equation (6.1) that with a given plant the noise levels at the output and plant input are decided once the decision concerning the form of T(s) has been made. It is therefore disadvantageous to select a system function of unnecessarily high bandwidth as this will only result in increased noise transmission and the likelihood of plant saturation.

In this context, Towill (5) has high-lighted the problem by pointing out that for a given system pole-zero geometry and a given plant, the tendency to saturation is increased by the ratio λ^e if the bandwidth is increased by λ, where e is the pole-zero excess of the closed-loop system. Additionally, the tendency to saturation as a result of the application of step inputs or high-frequency noise will be reduced by selecting a system function with a small ratio of the product of system poles to the product of system zeros. An important practical consequence is that a given bandwidth is achieved with a small pole-zero ratio by choosing an approximate Butterworth design of appropriate order.

6.3. OPTIMIZATION OF THE SYSTEM PARAMETERS

The use of tables of standard integrals of the previous section can be extended to simple optimization problems where the structure of the system has been decided in advance and all that remains is to select the most appropriate values for each of the parameters. Any given set of parameters will cause the system to transmit a certain amount of the signal noise whilst simultaneously causing some distortion of the true message or command signal. An optimum set of parameters will therefore result in the minimum mean-square error.

One disadvantage of the method is that the command signal must be given a spectral representation in order to compare it with the noise. If the assumption is made that the message and noise are uncorrelated, which is quite often justified, then the transmission of each may be handled separately and the total mean-square error is simply the sum of their individual components.

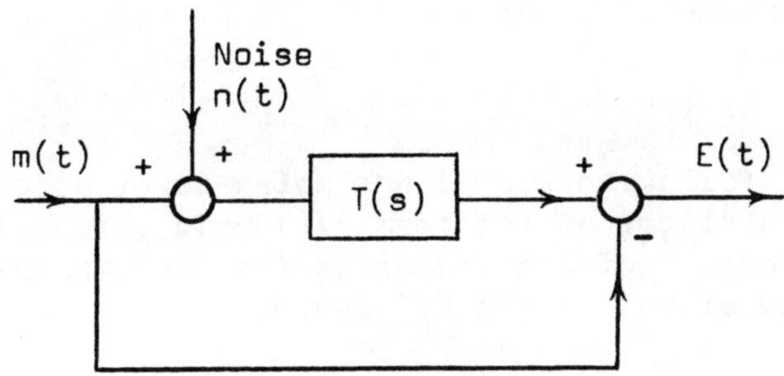

Fig (6.2): The total mean-square deviation due to noise and filter distortion of message m(t) is found by the addition of uncorrelated components.

Thus, if the system function is T(s) and the power spectral functions of message and noise are Φ_m and Φ_n respectively, then:

(a) Noise contribution at output:

$$\overline{e_n^2(t)} = \frac{1}{2\pi j} \int_{-j\infty}^{j\infty} T(s)T(-s)\Phi_n(s)ds \qquad 6.7a$$

(b) Message distortion:

$$\overline{e_m^2(t)} = \frac{1}{2\pi j} \int_{-j\infty}^{j\infty} \left[1-T(s)\right]\cdot\left[1-T(-s)\right]\Phi_m(s)ds \qquad 6.7b$$

The two components are found from the tabulated expressions in terms of the adjustable parameters and the total mean-square error or deviation at the output is then given by:

$$\overline{e^2(t)} = \overline{e_m^2(t)} + \overline{e_n^2(t)} \qquad 6.8$$

Equation (6.8) is differentiated partially with respect to each adjustable parameter in turn and equated to zero to give a set of simultaneous equations which are then solved to provide the optimum set to minimise the deviation.

Example

Given the following signal and noise power spectra and assuming a filter of the form $1/(1 + Ts)$, find T such that the output of the filter is the best approximation to the signal in the MSE sense.

$$\Phi_m(s) = \frac{1}{1 - s^2} \quad : \quad \Phi_n(s) = \frac{s^2}{s^2 - 1}$$

Thus, for the noise:

$$\overline{e_n^2(t)} = \frac{1}{2\pi j}\int_{-j\infty}^{j\infty} \frac{s}{(1+Ts)(1+s)} \cdot \frac{-s}{(1-Ts)(1-s)}\, ds$$

$$= \frac{1}{2T(1 + T)}$$

While for the signal:

$$\overline{e_m^2(t)} = \frac{1}{2\pi j}\int_{-j\infty}^{j\infty} \frac{Ts}{(1+Ts)(1+s)} \cdot \frac{-Ts}{(1-Ts)(1-s)}\, ds$$

Since the signal and noise are uncorrelated:

$$\text{Total MSE} = \frac{1 + T^2}{2T(1 + T)}$$

The total MSE is now minimised with respect to the filter time constant T:

$$\frac{d(MSE)}{dT} = \frac{2T^2 - 2 - 4T}{4T^2(1 + T)^2} = 0$$

On solving, we find the optimum time constant to be:

$$T_{opt} = 2.414 \text{ sec}$$

For a wider appreciation of the potential of statistical methods in control system design, the reader is referred to the comprehensive text by Newton, Gould and Kaiser (66).

6.4. DETERMINATION OF MSE VIA SIMULATION

Let H(s) be the transfer function of a linear system relating the input x(t) to the system error e(t).

(a) If x(t) is a stochastic process:

$$\overline{e^2(t)} = \frac{1}{2\pi j}\int_{-j\infty}^{j\infty} \Phi_e(s)ds$$

$$= \frac{1}{2\pi j}\int_{-j\infty}^{j\infty} H(s)H(-s)\Phi_x(s)ds \qquad 6.9$$

where Φ_x and Φ_e are input and error spectra.

(b) If x(t) is a transient starting at t = 0:

$$\int_0^{\infty} e^2(t)dt = \frac{1}{2\pi j}\int_{-j\infty}^{j\infty} E(s)E(-s)ds$$

$$= \frac{1}{2\pi j}\int_{-j\infty}^{j\infty} X(s)H(s).X(-s)H(-s)ds \qquad 6.10$$

Equation (6.10) is a consequence of Parseval's Theorem (Appendix D) and where X(s) and E(s) are the Laplace transforms of the input transient and the subsequent error.

On comparing equations (6.9,10) we see that if:

$$\Phi_x(s) = X(s)X(-s) \qquad 6.11$$

... <u>then the MSE arising in any linear system as a result of a random input x(t) is the same as the integral of the error squared due to the transient which fulfils the relationship (6.11)</u>.

This result (73) can be useful in simulation since it may be easier to apply certain transients to the system than to apply a random signal over and over again to ensure consistent results.

<u>Example</u>

A random process has been identified to have the spectrum below:

$$\Phi_x(j\omega) = \frac{2a}{(a^2 + \omega^2)}, \text{ or } \Phi_x(s) = \frac{2a}{(a^2 - s^2)}$$

Thus,

$$\Phi_x(s) = \frac{2a}{(a + s)(a - s)} = X(s)X(-s)$$

That is:

$$X(s) = \frac{(2a)^{\frac{1}{2}}}{(a + s)}$$

The equivalent transient is therefore given by:

$$x(t) = (2a)^{\frac{1}{2}}.e^{-at} \quad : \quad t > 0$$

The signal x(t) is therefore simulated, applied to the system and the resulting error is squared and integrated for a time which is sufficient for adequate convergence to occur.

6.5. THE SYSTEM AS A WIENER FILTER

There are some practical situations, notably in the area of weapon guidance, where it is imperative that the controlled system should follow the command signal with the least possible error despite the presence of heavy noise contamination of this signal. Such might be the weighting given to the importance of optimal following in comparison to other factors (at least in the initial investigative design studies) that the structure of the system function is no longer constrained but rather to be determined solely by the requirements of optimality.

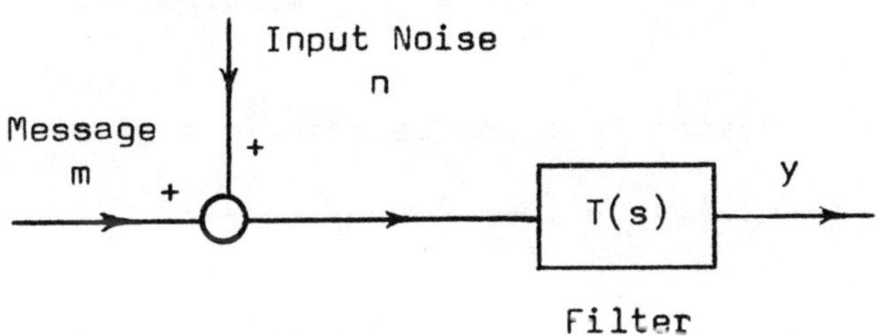

Fig (6.3): The closed-loop system T(s) considered as a filter to minimise the deviation between m and y.

From equations (6.7a,b) we need to minimise the mean-square error:

$$\overline{e^2(t)} = \frac{1}{2\pi j}\int_{-j\infty}^{j\infty}\{|T-1|^2\Phi_m + |T|^2\Phi_n\}ds \qquad 6.12$$

The function T(s) which minimises equation (6.12) can be found by making the subsitution $T(jw) = A\exp(j\theta)$ then initially selecting θ to effect a minimum $\overline{e}^2$ followed by a minimization with respect to the argument A. When this is done, it is found that the optimum system function $T_o(j\omega)$ is expressed simply as:

$$T_o(s) = \frac{\Phi_m}{\Phi_m + \Phi_n} \qquad 6.13$$

Now since the spectral density functions are restricted to rational functions in s^2 (they are assumed derived by operating on white noise with linear filters) the zeros of the denominator of T_o (ie its poles) will be distributed aymmetrically with respect to the $j\omega$-axis. That is, half the poles will lie within the RHP.

As an example, let:

$$\Phi_m(s) = \frac{1}{(1 + s)(1 - s)}$$

$$\Phi_n(s) = 1, \quad \text{ie white noise.}$$

The optimal system function is therefore given by:

$$T_o(s) = \frac{1}{(\sqrt{2} + s)(\sqrt{2} - s)}$$

Now a function as the one above, with RHP poles, may represent an unstable system or one corresponding to an impulse response which exists for negative time. The actual nature of the problem must be investigated to determine which interpretation is applicable. Thus, for the problem in hand, if the unstable interpretation were to be adopted, then $\overline{e}^2$ would be infinite and the minimization requirement would be invalidated. The optimal system thus turns out to be one which should have an impulse response existing over all time and therefore one which should respond to stimuli even before the stimuli have been applied and as such is therefore physically unrealizable.

This situation arises as a consequence of the unrealistic manner in which the minimization of the mean-square value was carried out, that is, by the independent minimization with respect to phase and magnitude of T(s) when in fact they are interdependent for real systems (see Appendix C).

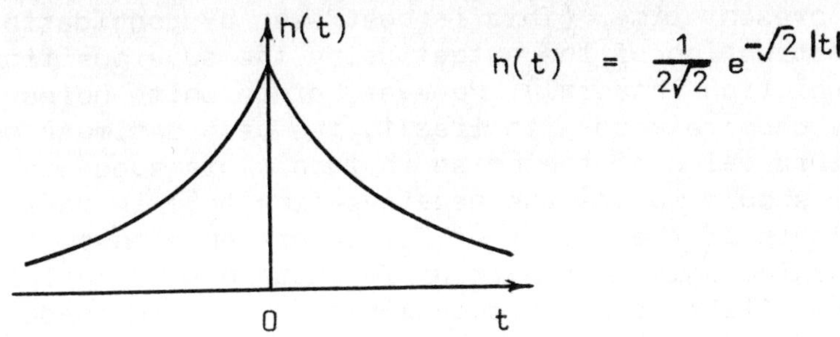

Fig (6.4): The impulse response h(t) of the optimal filter T_o for the example on p 138.

Wiener (70) was the first to contribute significantly in this area although it is to Bode and Shannon (71) that the simplified derivation is attributed. The principle in arriving at a realistic filter is to formulate T_o into two factored sections, T_1 and T_2, where:

$$T_1 = \frac{1}{(\Phi_m + \Phi_n)^+} \quad 6.14$$

where $(\ )^+$ indicates the factor of ( ) retaining all poles in the LHP.

The spectral density of the output of the first section is therefore given by:

$$|T_1|^2(\Phi_m + \Phi_n) = 1 \quad 6.15$$

That is, the output of T_1 is white noise which is then applied to T_2 which must have the form:

$$T_2 = \frac{\Phi_m}{(\Phi_m + \Phi_n)^-} \quad 6.16$$

T_2 is seen to have poles in both the LHP and RHP, the latter of which give rise to the negative-time tail of the impulse response which will cause future values of the white noise input to T_2 to contribute to its output at the present time. (This is best seen by considering the construction of the output using the superposition or convolution integral). However, since white noise is totally uncorrelated with itself, the best estimate of any future value of the noise is zero. This suggests that we should ignore the negative-time tail in making an estimate of the output of T_2. We are able to do this by retaining only that part of T_2 which has LHP poles. The final filter therefore consists of T_1 in cascade with the physically realizable portion of T_2 to give the physically realizable and optimal Wiener filter T_o.

$$T_o = \frac{1}{(\Phi_m + \Phi_n)^+} \cdot \left[\frac{\Phi_m}{(\Phi_m + \Phi_n)^-}\right]^+ \qquad 6.17$$

where $[\ ]^+$ indicates the contribution (partially fractioned) of the LHP poles.

Using the above concepts to arrive at the Wiener filter form of the optimum system function for the example of p 138, we have:

$$T_1 = \frac{(1 + s)}{(\sqrt{2} + s)}$$

$$T_2 = \left[\frac{1}{(1 + s)(\sqrt{2} - s)}\right]^+$$

$$= \frac{1}{(1 + \sqrt{2})} \cdot \frac{1}{(1 + s)}$$

The optimum system is therefore given by:

$$T_o = \frac{1}{(1 + \sqrt{2})} \cdot \frac{1}{(\sqrt{2} + s)}$$

Note the dependence of the order of T_o upon the orders of the signal and noise models. The reader is referred to alternative texts (32,68,69) for further consideration and development of the above material.

6.6. EXTERNAL NOISE DISTURBANCE

The external disturbances which influence a system may be regarded as plant inputs which are additional to those which command the behaviour of the system and which arise as a result of changes in the environment of the plant or in changes of status of other systems coupled to the plant. Obvious examples include variations of demand on power-supply systems, wind effects on scanning antennas and the influence of sea-state on platform stabilization.

It was seen, when discussing the feedback principle, that an adequate loop function would enable reduction of the effects of a disturbance to a suitable level. However, the common feature of plant variation adds to the difficulty of the problem so that if the designer were ever able to measure the disturbances he would still not be in a position to effect a cancellation by use of 'disturbance feedforward' as has been suggested.

The design problems introduced by plant variation and external disturbances are intimately related, as we have seen, via the loop function and it is to be expected that the frequency-domain method of Horowitz which has been expounded in the context of sensitivity reduction should be equally applicable to the rejection of noisy or deterministic disturbances acting on the plant. This is indeed the case.

It will be assumed that the disturbance enters the system at the input to the plant and not at some intermediate point within the plant since only a trivial modification is necessary to the development to handle other situations. The disturbance transmission function is therefore given by:

$$T_d(s) = \frac{P(s)}{1 + P(s)H(s)G(s)} \qquad 6.18$$

where the functions P(s),H(s) and G(s) refer to the plant, feedback and feed-forward functions respectively.

On introducing the nominal plant function $P_o(s)$, the previous equation can be manipulated as follows:

$$T_d(s) = \frac{P(s)}{1 + L_o(s)\frac{P(s)}{P_o(s)}}$$

$$= \frac{P_o(s)}{\frac{P_o}{P}(s) + L_o(s)} \qquad 6.19$$

where L_o is the nominal loop function and P is any representative of the variable plant.

Equation (6.19) has the simple graphical interpretation of figure (6.5).

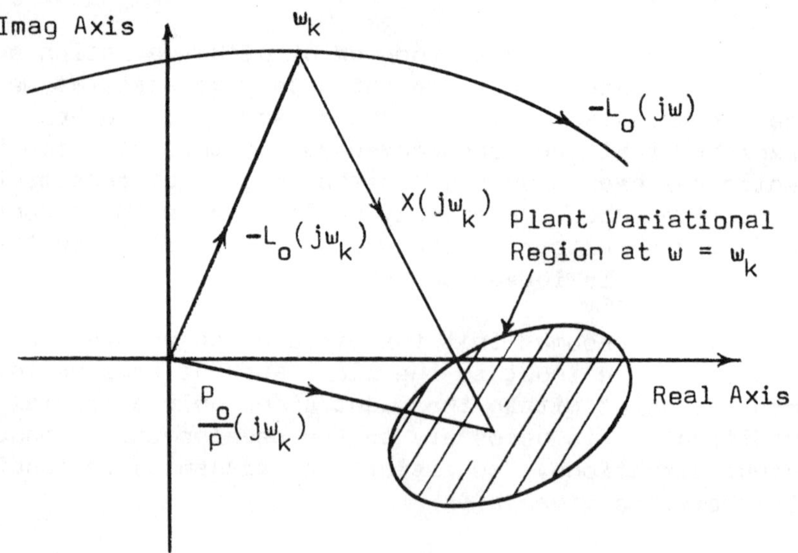

Fig (6.5): Graphical interpretation of equation (6.19), applicable at the frequency ω_k.

It will be seen that the vector $X(j\omega_k)$ may be expressed as follows:

$$X(j\omega_k) = \frac{P_o}{P}(j\omega_k) + L_o(j\omega_k) = \frac{P_o}{T_d}(j\omega_k) \qquad 6.20$$

Thus, given the nominal system loop function and the range of variation of the plant over all significant frequencies, the range of variation of the disturbance transmission function can be obtained by purely graphical means. Alternatively, should a desired transmission function be defined either from the spectral properties or the deterministic form of the expected disturbances, then the required loop function to ensure the specified transmission is known. Using the latter approach to loop synthesis, $X(j\omega)$ is specified over a range of frequencies and a set of circles is drawn at each frequency, with one circle for each possible plant variant centred on P_o/P and of radius mag$X(j\omega)$. The outer envelope containing all circles at ω_k will therefore be the boundary of mag$L_o(j\omega_k)$ within which the loop function must not encroach if the resulting disturbance transmission function is to have a magnitude equal to or less than the specified $T_d(j\omega_k)$.

Perhaps the least straightforward part of the above design procedure lies in the choice which must be made concerning the required form of the transmission function. One is generally not too concerned about the exact shape of the response to a deterministic disturbance provided it does not exceed certain maximum values and decays sufficiently quickly to zero or some small constant value. When the disturbance is stochastic, the method of section (6.2) can be used to estimate the MSE due to the disturbance (which requires an estimate of the spectrum of the disturbance) on the assumption of a given form of T_d, bearing in mind the plant function and the compensation complexities which can arise with an inappropriate or unrealistic choice of T_d.

Since the satisfaction of deterministic-response specifications are often more exacting, these will be considered for the remainder of this section. We decide upon an acceptable disturbance response and associate its characteristics with the coefficients of a model T_d. For example, an acceptable response to a unit step disturbance might be as shown in figure (6.6).

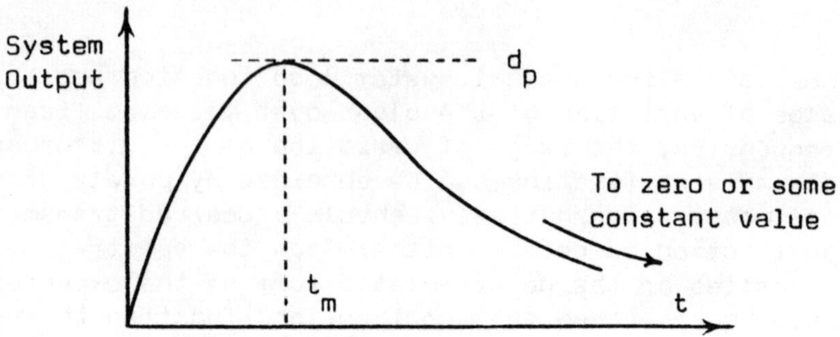

Fig (6.6): Envelope of acceptable disturbance responses.

The simplest transfer functions which produce the response of figure (6.6) when excited by step functions are:

$$T_d(s) = \frac{Ks}{(1 + as)^2} \quad : \text{ for zero SS response.}$$

$$T_d(s) = \frac{K(1 + bs)}{(1 + as)^2} \quad : \text{ for constant SS response.} \tag{6.21}$$

The above transfer functions are very simple and would not be sought in a real system since the loop function would need to be unnecessarily complex. However, they may be interpreted as guideline envelopes which can of course be oscillatory. It should be noted that if a zero steady-state response is demanded, the implication is that one additional integration will be needed in the loop beyond those already included in the plant itself. Further, the shorter the time to the peak of the response the greater will be the loop bandwidth with all the attendant noise problems.
As with the loop design to achieve sensitivity reduct-

ion, the loop-design procedure to ensure adequate disturbance rejection can become quite laborious if it is pursued in a graphical manner. The software requirement for both procedures to be carried on with the aid of a computer are very similar and are best incorporated in one CAD suite for convenience. The essential routines have been discussed previously under sensitivity reduction and will not be elaborated further. However, an example might reinforce the discussion so far.

Example

It is required to determine the loop function to employ with the plant defined below such that it has the minimum magnitude at all frequencies (without resorting to extensive compensation) and yet which ensures that the response of the system to a unit step disturbance entering at the plant input has a maximum amplitude d_p of 0.2 units at a time 0.5s after the application of the disturbance and which ultimately decays to a steady state value of 0.02.

$$P_o(s) = \frac{K(1 + 0.5s)}{s(1 + s)}$$

where K_o = 1 and all three parameters have a 50% tolerance.

The computer routine selects the second expression of equation (6.21) as a model for which the step disturbance response is:

$$d(t) = \overline{K}(1 - e^{-t/a}) + \frac{\overline{K}(b - a)}{a^2}.te^{-t/a}$$

where a = 1.0, b = 0.5 and $\overline{K}$ = 0.02

Trivial manipulation leads to:

$$d_p = \overline{K} + \frac{\overline{K}b}{a}\left(\frac{b - a}{b}\right).e^{-\left(\frac{b}{b - a}\right)}$$

$$t_m = \frac{ab}{b - a}$$

On providing the design data t_m, d_p and $\overline{K}$, the parameters of the model can be determined in an iterative manner using Newton's algorithm and the radius vector calculated at every frequency of interest. The first frequency is now selected and the boundary of 1000 plant variants is computed as for the sensitivity reduction programme, from which is obtained the envelope of all circles of radius $X(\omega_1)$. This is the loop-magnitude boundary at ω_1. Repetition of this procedure yields the complete set of loop-magnitude boundaries after which the plant frequency response is superimposed and compensation incorporated until the loop response satisfies the boundary constraints.

One satisfactory compensator for this example is found to be:

$$G(s)H(s) = \frac{50(1 + 0.7s)}{(1 + 6.0s)}$$

The design of the loop function is complete, leaving the distribution of this function between G(s) and H(s) to be decided by the dynamic response required of the closed-loop system. Figure (6.7) shows the computer display of some boundaries together with a portion of the loop frequency response, from which it is seen that there is some gain deficiency in the region of 1.5 - 3.0 rad/s. This can be ignored initially but corrected later if necessary on examination of the simulated responses. These are also illustrated in figure (6.7) indicating a performance well within the demanded specifications.

6.7. FEEDBACK NOISE

Noise of any significant level which is introduced into the feedback path can be a source of serious problems when the closed-loop system has been designed in a realistic manner to account for non-ideal hardware. In

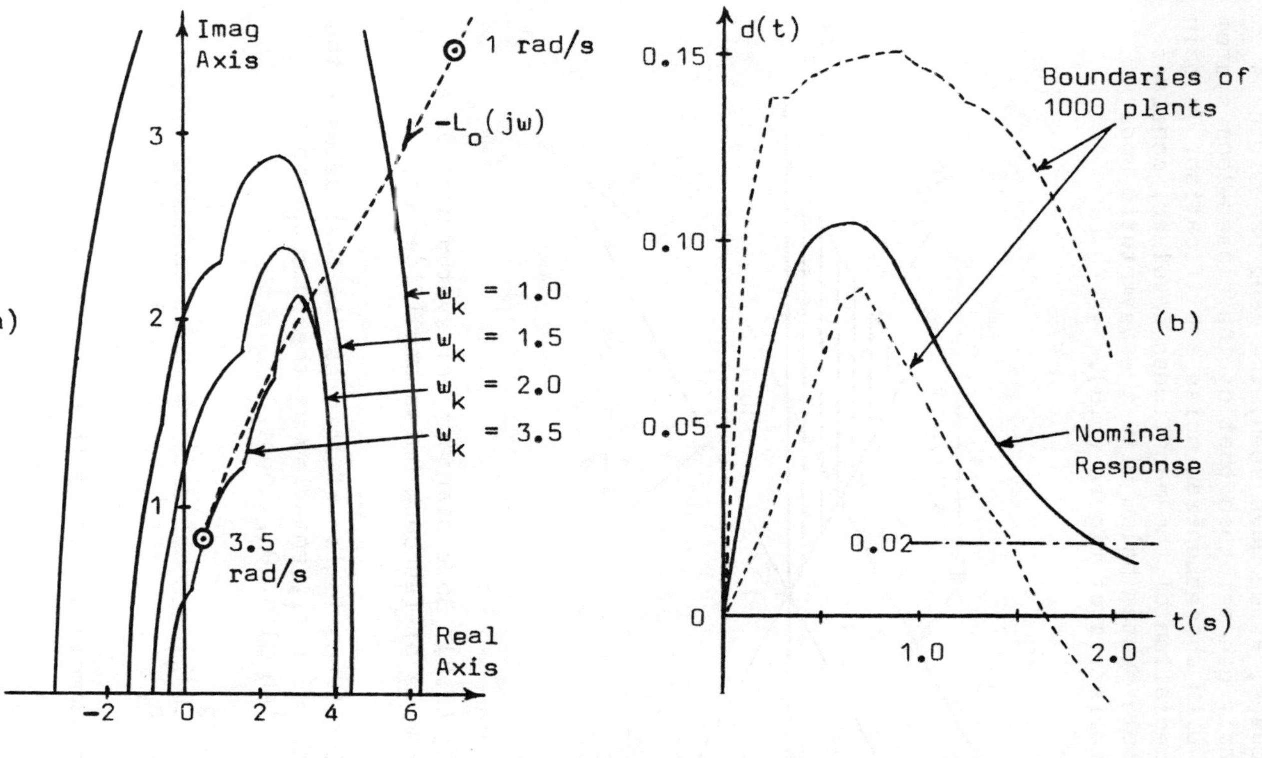

Fig (6.7): Results for the compensated system showing (a) a number of loop-magnitude boundaries for $-L_o$ and (b) the nominal disturbance response within the envelope of responses of 1000 plant variants.

these cases, as we have seen, the loop bandwidth is inevitably greater than that of the closed-loop system itself with consequential noise amplification. To gain an appreciation of the magnitudes involved, consider the simple system below with the asymptotic Bode characteristics of figure (6.8).

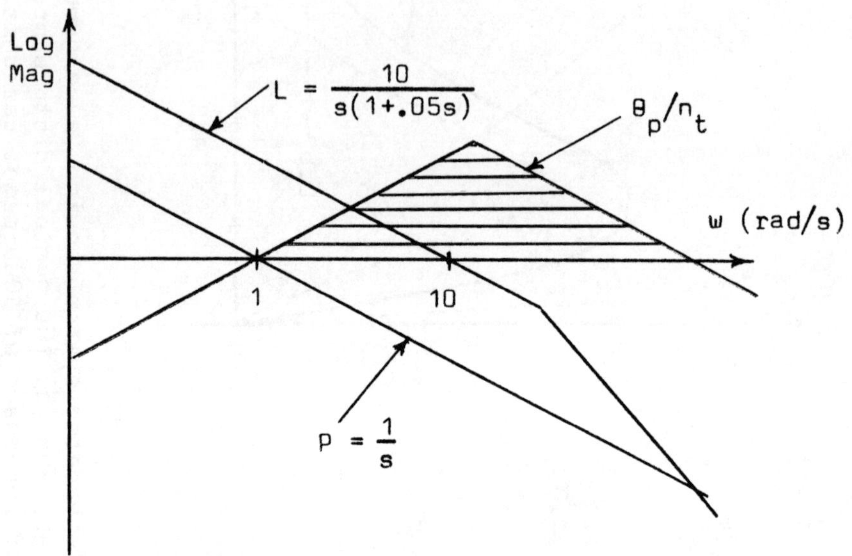

Fig (6.8): Bode diagrams for system used as an example of feedback noise amplification.

The noise transmission of interest is that between the feedback noise n_t, entering at the output, and the input to the plant θ_p, thus:

$$\frac{\theta_p}{n_t}(s) = \frac{G(s)H(s)}{1 + L(s)} = \frac{1}{P(s)} \frac{L(s)}{1 + L(s)}$$

For the example of figure (6.8), we have:

$$\frac{\theta_p}{n_t}(s) = \frac{s}{1 + 0.1s + 0.005s^2}$$

The mean-square plant input due to noise is therefore:

$$\overline{\theta_p^2(t)} = \frac{c_0^2 d_2 + c_1^2 d_0}{2d_0 d_1 d_2} = 1000$$

.. if, $\Phi_{nt} = 1$

Thus, with a not over ambitious loop function and a unit spectral density noise source, approximated here by white noise, there is a power amplification of one thousandfold. The area of amplification is shown shaded in figure (6.8) and as can be seen may extend over a considerable frequency range well beyond the system bandwidth.

Since the primary source of feedback noise lies in the measurement transducers used, the penalty for the use of feedback may manifest itself purely in financial terms since critical problems may demand only the beat hardware. As an example of the magnitude of transducer noise in comparison with the actual changes of the signal which may be experienced in practice, figure (6.9) shows the feedback signal obtained from a superheated steam-flow meter of a marine boiler (72)

Alternatively, significant noise levels may be experienced directly at the output of some systems such as aircraft in turbulence and ships in heavy seas. Unfortunately these latter noise sources are unavoidable and often of huge proportions such that, with the wide bandwidths necessary to minimise the effects of the massive parametric variation exhibited by these vehicles, it is often necessary to employ multi-mode operation permitting selective bandwidth according to conditions. In this situation, the designer can be faced with the dilemma of authorative control on the one hand and hardware wear-and-tear and power loss on the other.

Figure (6.10) shows the response of the USS Compass Island, a Mariner-Class marine vessel, to a course-change demand of 0.5 radian with a very low level sea-state disturbance of 0.6^0 peak-to-peak causing

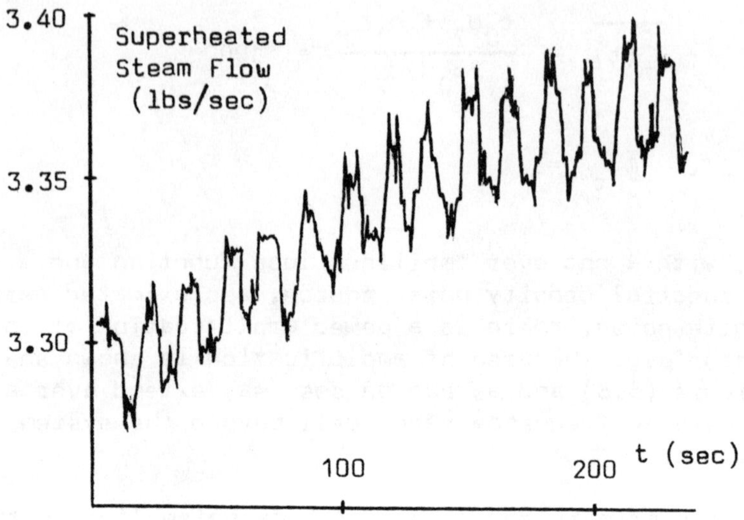

Fig (6.9): Variation of superheated steam-flow rate of a marine boiler in response to a 11% increase in fuel-flow rate.

the slight yaw deviation in course (53). With a controller to reduce parametric variation of the ship of up to 300%, the system bandwidth spans the entire noise spectrum giving rise to inordinate rudder activity as illustrated in the figure. The mean rudder demand to effect the course change is just discernable.

There is a limit therefore to what can be achieved with a linear time-invariant two degree-of-freedom system, the restraining factor being the level of feedback noise. As suggested, multi-mode operation may be contemplated with the bandwidth selected according to the prevailing conditions. This has certainly been applied to aircraft and to the manoeuvring control of VLCC vessels to avoid the difficulties of figure (6.10). Another approach which has been investigated with respect to ships is the continuous identification via a recursive least-squares algorithm of the changing dynamics of the ship during a manoeuvre (75). The aim was to reduce the range of uncertainty thus relieving the controller of the need for a very wide bandwidth. It was found however that the dynamic variation was

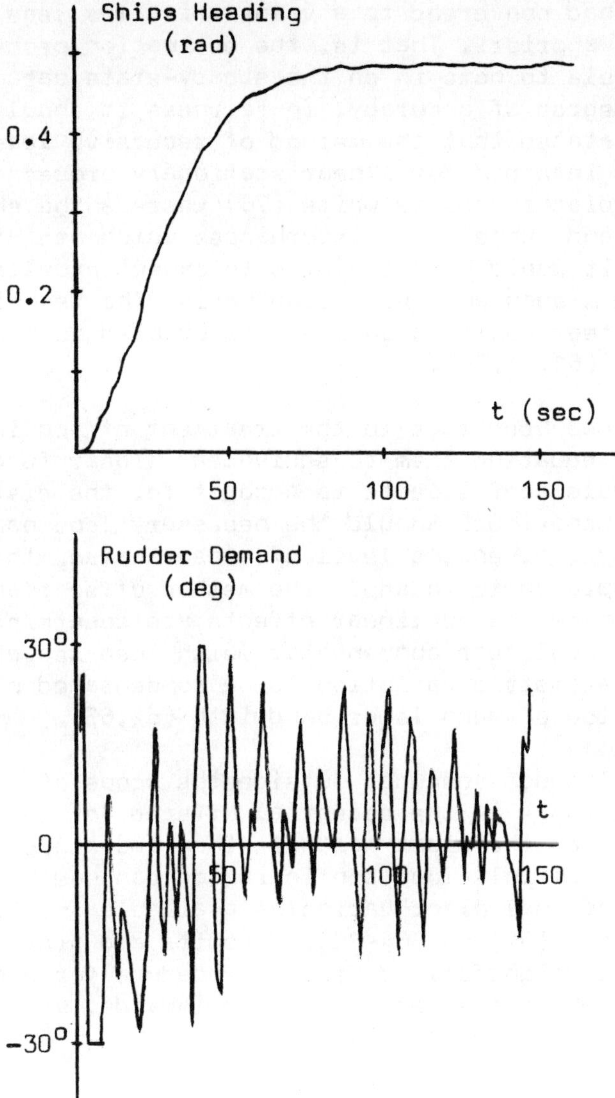

Fig (6.10): Ships heading and rudder demand for a 0.5 radian course-change demand of the USS Compass Island. Although the effect of yaw disturbance is slight on the actual course, the high bandwidth controller induces intense rudder activity.

substantially complete before the variance of the RLS estimates had converged to a value which was less than that known a priori. That is, the estimation process was only able to home-in on the steady-state parameters with any degree of accuracy. In fairness it should of course be stated that the method of recursive least-squares is intended for linear stationary processes for which the disturbance is white (76) whereas the ship is nonlinear and subject to disturbances which are highly coloured. It would be optimistic to expect excellent results from such an unpromising basis. The technique finds greatest application when the problem is one of regulation (61,77,78).

Reference has been made to the treatment of nonlinear systems by equating them to equivalent linear functions with variable coefficients to account for the distinctly nonlinear behaviour. Should the necessary loop bandwidth be excessive, given the level of sensor noise, then the obvious approach is to apply the method of section (5.16.2) where the nonlinear effects are countered by an inverse nonlinear compensator which then leaves only the true parametric variation to be compensated by a loop function of much lower bandwidth (62,63).

Finally, although somewhat outside the scope of this monograph, there is the potential offered for noise reduction when the plant has more than two degrees-of-freedom. A multiple-loop configuration can now be used with the internal plant variables available for feedback purposes (32, pp 364-83). Horowitz and Sidi (57) present a straightforward design procedure for such systems in which the designer works inwards, step by step, from the outermost loop.

6.8. DISTURBANCE CONTROL VIA THE NICHOLS CHART

Should it be the intention of the designer to pursue system synthesis in the manner of Section (5.13) then a novel means of restricting the magnitude of disturbance transients has been provided by Sidi (48,54). For his formulation it is necessary to assume that the disturbance enters directly at the output such that the disturbance transmission function is now given by:

$$T_d(s) = \frac{1}{1 + L(s)} \qquad 6.22$$

It is important to note that the function $T_d(s)$ is identical to the sensitivity function as originally defined. The logarithmic magnitude changes of equations (5.19,20) are however unaltered by the relocation of the disturbance.

Now it was pointed out in Section (4.3) that for any realistic system there must be as much positive as negative feedback over the whole frequency range. Thus with the potential for sensitivity reduction being concentrated at the lower frequency range, there must be some frequency above which the sensitivity function becomes greater than unity. In the context of sensitivity reduction this is of secondary importance provided that frequency were well beyond the critical range for the system. That is, the system function would be so small that even gross percentage changes would be of no significance to its behaviour. Unfortunately this is not so when one consider disturbance response as defined by equation (6.22). An increasing sensitivity function implies an increasing transmission of disturbance.

This being the case, how is one to curtail disturbance response in the design process when using the Nichols chart? Sidi introduced the transformation $m = 1/L$ such that:

$$T_d(s) = \frac{m(s)}{1 + m(s)} \qquad 6.23$$

Equation (6.23) is now in the correct form for use on the chart after allowance has been made for the transformation by reflecting the phase-angle about the -180° axis and the gain about the Odb axis. (Actually, since we are concerned with magnitude boundaries, only the latter reflection need be made). Sidi's approach was now to equate the characteristic equation of equation (6.22) to a second-order polynomial and to correlate the peak amplitude-ratio of the frequency response to the damping factor of the transient response. Thus, a peak of 8db would correspond to a damping ratio of 0.2, whilst a 2.7db peak would similarly give rise to a damping factor of 0.4. Such a specification would there fore define a region in the centre of the Nichols chart through which the loop function must not pass.

For example, consider the boundaries illustrated in figure (6.11).

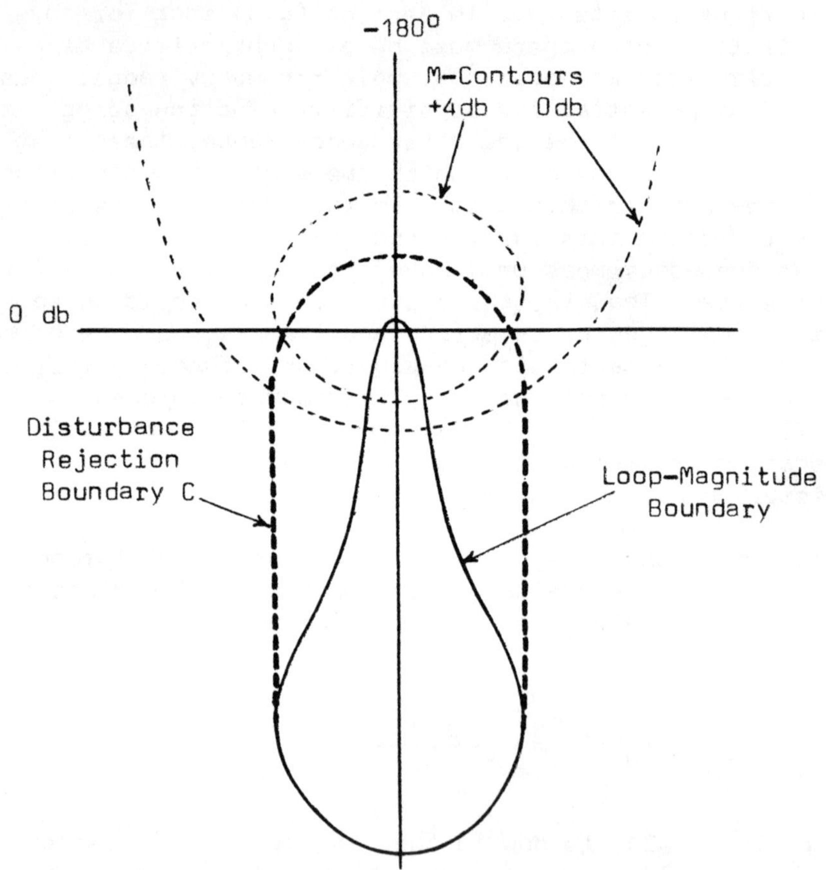

Fig (6.11): Construction of boundaries on the Nichols chart to restrain the damping factor of the disturbance response.

The loop-magnitude boundary has been drawn for a frequency beyond the sensitivity function crossover at which there is negligible phase difference between the plant variants and for which $\Delta|L| = \Delta|P| = 20$db and $\Delta|T| = 23$db. Suppose now that the maximum peak resonance of T_d is to be +4db. The +4db M-contour is reflected about the 0db axis and, to ensure that no variation of $P(j\omega)$ will cause the loop function to enter the inverted region, vertical boundaries are dropped 20db or until they contact the loop-magnitude boundary. The contour C therefore defines the region through which the loop function must not pass if <u>both</u> sensitivity and disturbance rejection specifications are to be met.

CHAPTER 7

Constraining the Control Effort

7.1. INTRODUCTION

It is fortunate that one aspect of the human condition is an urge always to strive for excellence in whatever endeavour the species is implicated, only deviating from this inveterate objective when circumstances, means, knowledge or the penalty of cost or effort give birth to the inevitable compromise. In no other field is the conflict of best and possible and the dilemmas of compromise better illustrated than in engineering where real physical processes and not hypothetical models are the order of the day. Such is the case of course in control engineering where the diabolical influence is ignorance and feedback provides the saving grace.

Earlier in the history of the subject engineers were, in view of the state of the art and facilities which were available, quite satisfied with a definition of best which gave primary consideration to relative stability and steady-state accuracy. Criteria of performance such as gain and phase margins were in vogue and a heavy dependence was placed on intuitive notions of transient responses and system resonance. Later, attempts were made to quantify the 'goodness' of the control system by the setting up of performance criteria with an analytic base, some of which still provide useful guidelines to this day, but still with an eye to dynamic accuracy. The search for a measure of perfection has continued but has foundered to some extent on mathematical tractibility. Indeed, there is as yet no meaningful definition of optimum which permits an analytic solution for other than trivial cases. The subject of optimization has necessarily involved the use of computational facilities of one kind or another for all real problems. Even those techniques with limited performance criteria which admit of an analytic solution require the assistance of the computer in translating them into useable information.

Before continuing with a discussion on the means of assessing performance, it is as well to give some thought to the meaning which can be attached to the word 'optimal' which is used freely in the literature. In contrast to everyday usage implying the best possible or the most excellent, the term has, in control engineering, a very particular meaning. This restricted definition is that an optimum system is one which is optimal with respect to one particular performance index which is itself weighted according to the dictates of the user. It can therefore be a very subjective index with definition and weighting decided by the intuition of the designer, perhaps with his eye on the response guidelines used by his fore-runners.

7.2. PERFORMANCE INDICES

A desirable feature of any control system is that it should function with a satisfactory degree of accuracy, both steady-state and dynamic, thus calling for a specification of the maximum allowable error from all sources when responding to deterministic inputs. However, in place of this relatively simple demand, a performance index may be substituted which has the advantage of encompassing in a single number a measure of the ability of the system to maintain a low level of error. Very often the performance index J is defined as:

$$J = \int_0^\infty f(t,\theta_e)dt \qquad 7.1$$

where θ_e is the system error and time may be included to penalise persistent errors.

The three most common indices of this type are defined in expression (7.2) of which only (7.2b) is amenable to analytic solution.

(a) Integral of Absolute Error:

$$J = \int_0^{\infty} |\theta_e| \, dt \qquad 7.2(a)$$

(b) Integral of Error Squared:

$$J = \int_0^{\infty} \theta_e^2 \, dt \qquad 7.2(b)$$

(c) Integral of Time x Absolute Error:

$$J = \int_0^{\infty} t|\theta_e| \, dt \qquad 7.2(c)$$

In using the above indices, the objective of the designer is to minimise the integral by a suitable selection of the values of the adjustable parameters of the system. To do this however for the Integral of Absolute Error (IAE) or the Integral of Time x Absolute Error (ITAE) would involve the user in an optimization exercise requiring analogue or digital simulation.

The IAE index is seen to penalise any error and its use often results in a system which is relatively overdamped, whereas the IES bears down heavily on large errors, biasing towards low overshoot. By concentrating on error magnitude, the latter has a tendency to produce longer settling times and is associated with oscillatory responses. Persistent transients, farcical in a grammatical sense but factual nevertheless, can be penalised by the inclusion of time as a factor of the integrand as with the ITAE index, with correspondingly less emphasis on the reduction of the initial overshoot. An advantage of this latter index is its enhanced selectivity in that the minimum curve of J is more narrow, resulting in large changes of J for any deviation from the optimum set of parameters. Generally, it gives rise to an acceptable form of response (79).

Although the judicious use of performance indices can result in satisfactory system behaviour it is nevertheless true that particular aspects of the response, which may be of some importance to the functioning of the system, are out of the control of the designer. Other than the more obvious transient overshoot and resonance peak, he may be interested in transmission of noise, ramp-peak error and the level of demand made upon the plant to name but a few. Consideration of such criteria and their inclusion within a performance index has led of late to the discrete index defined by equation (7.3), (80,81,27).

$$J = \sum_{i=1}^{m} \left(1 - \frac{x_i}{x_{id}}\right)^2 \qquad 7.3$$

where x_i ~~is~~ the ith criterion with the current configuration and x_{id} is the value demanded and where the summation is over the m individual criteria.

There are two immediate difficulties with this latter index: firstly, the labour of deriving expressions for the x_i for each design problem is excessive and, secondly, the expression for the index will become transcendental. The first difficulty can be overcome by designing towards dominant models of the required system (9,82), where criteria values are conveniently computed as functions of the model parameters. (It is found in practice that certain dominant models can represent a wide range of systems requiring therefore the once-for-all computation of one set of normalized design curves or computer files). The second difficulty that of the minimization of the performance index, is eased only by resorting to the use of a computer. A particularly simple method of minimization, both conceptually and from the point of view of software development, which the author has found to be very useful, is a version of the Simplex method as employed by Nelder and Mead (8) which does not involve the determination of gradients and which is suitable for use with functions which are nonlinear or which exhibit discontinuities.

7.3. AN INDEX TO CONSTRAIN THE CONTROL EFFORT

Most physical systems are, strictly speaking, nonlinear and of all the individual expressions of nonlinearity the one most often encountered with control systems is saturation. Sometimes it is the case that whilst endeavouring to achieve the design objectives, demands are made upon the plant which it is physically incapable of satisfying, leading to a deterioration of performance from that expected. In others, noise may be injected into the loop which, if amplified by any differentiation within the feedback path, can lead to saturation of the plant by noise. A more useful index of performance would therefore be that of equation 7.2b subject to the condition:

$$\text{Max } |u(t)| \leqslant K \qquad 7.4$$

where $u(t)$ is the plant input.

However the invocation of the above condition can result in a nonlinear controller. If one wishes to constrain the level of plant input whilst still retaining a linear and mathematically soluble performance index, the best that can be done is to employ the quadratic index of equation (7.5).

$$J = \int_0^\infty (\theta_e^2 + \lambda u^2)dt \qquad 7.5$$

where λ is a factor which weights the significance of error and level of plant input.

Obviously, the greater the value assumed by λ, the greater will be the degree of significance which is to be placed upon large excursions of the control effort. Having decided upon the weighting factor, the use of equation (7.5) will undoubtedly provide an 'optimal' controller, but only in the sense that equation (7.5) has been minimised. Another designer may have opted, according to his objectives, for an entirely different weighting.

Although not totally satisfactory, the quadratic index will be used in the remainder of this chapter to define the optimal controller. Initially however we must decide on the approach to be adopted, whether via the frequency or the time domain. In keeping with the philosophy of the monograph, the design method selected will aim for the specification of an optimal transfer function which, for a given plant, may be implemented in a manner which is at the discretion of the designer. The alternative approach via the time domain, which achieves its objectives by the definition of an optimal control law (ie, a statement of u(t) in terms of the state variables), is well covered in standard texts (47) and is reviewed briefly in Appendix (E).

7.4. TRANSFER-FUNCTION APPROACH TO OPTIMALITY

The performance index of equation (7.5) must be converted from its time-domain formulation into one involving the system transfer function, the plant and the system input. This adaptation is conveniently performed by means of Parseval's Theorem (see Appendix D) which expresses the relationship between a function x(t) and its Laplace transform X(s) as follows:

$$\int_0^\infty x^2(t)dt = \frac{1}{2\pi j}\int_{-j\infty}^{j\infty} X(s)X(-s)ds \qquad 7.6$$

We will use the following transfer functions relating the system error θ_e and plant input u to the system input θ_i:

$$\frac{\theta_e}{\theta_i}(s) = 1 - T(s) \; ; \; \frac{u}{\theta_i}(s) = T(s)/P(s) \qquad 7.7$$

where T,P represent the overall system function and plant respectively.

Now, using equation (7.6,7), the index J of equation (7.5) can be re-written as:

$$J = \frac{1}{2\pi j} \int_{-j\infty}^{j\infty} \left[1 - T(s)\right] \left[1 - T(-s)\right] \theta_i(s)\theta_i(-s) ds$$

$$+ \frac{\lambda^2}{2\pi j} \int_{-j\infty}^{j\infty} \frac{T(s)T(-s)\theta_i(s)\theta_i(-s)}{P(s)P(-s)} ds \qquad 7.8$$

The objective is to minimise J, for a given θ_i, by the selection of an optimum T(s). Now, following the approach of Chang (83), any arbitrary T(s) may be expressed as:

$$T(s) = T_o(s) + kT_1(s) \qquad 7.9$$

where T_o is that selection which minimises J and where k is an arbitrary constant.

On substitution of (7.9) into (7.8), we have:

$$J = J_1 - k(J_2 + J_3) + k^2 J_4 \qquad 7.10$$

The component indices of equation (7.10) are given in equation (7.11).

$$J_1 = \frac{1}{2\pi j}\int_{-j\infty}^{j\infty}\left\{\left[1 - T_o(s)\right]\left[1 - T_o(-s)\right] + \lambda^2 \frac{T_o(s)T_o(-s)}{P(s)P(-s)}\right\}\theta_i(s)\theta_i(-s)ds$$

$$J_2 = \frac{1}{2\pi j}\int_{-j\infty}^{j\infty}\left[-1 + T_o(s) + \lambda^2 \frac{T_o(s)}{P(s)P(-s)}\right]\theta_i(s)\theta_i(-s)T_1(-s)ds$$

$$J_3 = J_2(-s) \tag{7.11}$$

$$J_4 = \frac{1}{2\pi j}\int_{-j\infty}^{j\infty}\left[1 + \frac{\lambda^2}{P(s)P(-s)}\right]T_1(s)T_1(-s)\theta_i(s)\theta_i(-s)ds$$

We notice that by definition J_1 is the minimum value of J with T_o substituted for T and that J_4 is always positive. Therefore, for arbitrary T_1, the sufficient condition for T_o to give the lowest value of J is that $J_2 = J_3 = 0$. That is:

$$\int_{-j\infty}^{j\infty} \left[-1 + T_o(s) + \lambda^2 \frac{T_o(s)}{P(s)P(-s)}\right]\theta_i(s)\theta_i(-s)T_1(-s)ds = 0 \qquad \ldots . \quad 7.12$$

The form of equation (7.12), with the integration being carried out along the infinite imaginary axis, reminds one of a contour integration problem. Thus, if the contour is extended in an infinite semicircle and if the closed contour encircles no poles then the contour integral will vanish by the Residue Theorem. To treat equation (7.12) in this manner we must satisfy two conditions: namely, that the integrand vanishes as s approaches infinity (to allow integration only along the imaginary axis) and secondly that all the poles lie in one half-plane.

The latter is easily decided since as $T(s)$ represents a stable system all the poles of $T_1(-s)$ must lie in the RHP. Therefore, denoting the remainder of the integrand by $Z(s)$, we see that $Z(s)$ must also have all of its poles in the RHP to permit solution. To examine the behaviour of the integrand at infinity we notice that it is made up of products of θ_i, $T\theta_i = \theta_o$ and $T\theta_i/P = \theta_p$ where θ_o and θ_p represent system output and plant input. Now since these functions are either zero or finite for $t > 0$, their transforms decay as s approaches infinity by at least $1/s$. The integrand therefore vanishes along the semi-circular part of the contour.

We therefore have:

$$\left[T_o(s) - 1 + \lambda^2 \frac{T_o(s)}{P(s)P(-s)}\right]\theta_i(s)\theta_i(-s) = Z(s) \qquad \ldots\ldots \quad 7.13$$

Now the function $1 + \lambda^2/P(s)P(-s)$ is unaffected by a change in sign of s and can therefore be factored into two functions Y(s) and Y(-s) where the poles of the former lie in the LHP and those of the latter lie in the RHP. This action of factoring into two functions according to their pole-zero locations is termed spectral factorization. We can therefore write:

$$Y(s) = \left[1 + \frac{\lambda^2}{P(s)P(-s)}\right]^+ \qquad 7.14$$

The superscript + indicates that the function has all its poles in the LHP and that its time-domain equivalent exists only for positive time. Now inserting (7.14) into (7.13), we have:

$$-\frac{\theta_i(s)}{Y(-s)} + Y(s)\theta_i(s)T_o(s) = \frac{Z(s)}{Y(-s)\theta_i(-s)} \qquad 7.15$$

The function $\theta_i(s)/Y(-s)$ is now decomposed, via partial fraction expansion, into two parts having poles in the LHP and RHP respectively, indicated by + or - subscripts such that:

$$-\left[\frac{\theta_i(s)}{Y(-s)}\right]_+ + Y(s)\theta_i(s)T_o(s) = \frac{Z(s)}{Y(-s)\theta_i(-s)} + \left[\frac{\theta_i(s)}{Y(-s)}\right]_-$$

$$\ldots\ldots\ldots \quad 7.16$$

Since the left-hand side of equation (7.16) has all its poles in the LHP whilst that of the right-hand side has all its poles in the RHP, each side must be equal to a constant which is readily shown (83) to be zero. The expression which defines the optimal transfer function is therefore given by:

$$T_o(s) = \frac{1}{Y(s)\theta_i(s)}\left[\frac{\theta_i(s)}{Y(-s)}\right]_+ \qquad 7.17$$

We will derive particular expressions for the cases when θ_i represents a step and a ramp input.

(1) T(s) optimal with respect to a step input.

$$\theta_i(s) = \frac{1}{s}$$

$$T_o(s) = \frac{s}{Y(s)}\left[\frac{1}{sY(-s)}\right]_+ = \frac{s}{Y(s)}\frac{1}{Y(0)s}$$

Thus:

$$\boxed{T_o(s) = \frac{1}{Y(0)Y(s)}} \qquad 7.18$$

(2) T(s) optimal with respect to a ramp input.

$$\theta_i(s) = \frac{1}{s^2}$$

$$T_o(s) = \frac{s^2}{Y(s)}\left[\frac{1}{s^2 Y(-s)}\right]_+$$

$$= \frac{s^2}{Y(s)}\left[\frac{1}{s^2 Y(0)} + \frac{Y'(0)}{sY(0)^2}\right]$$

Thus,

$$\boxed{T_o(s) = \frac{Y(0) + sY'(0)}{Y(s)Y(0)^2}} \qquad 7.19$$

The expressions (7.18,19) are simplified even further if P(s) has a denominator factor s^n, $n \neq 0$, since then Y(0) has the value of unity.

A consequence of equation (7.14) is that the optimal system has the same order as the plant which, for the case of unity-feedback of the output, implies some cancellation of the plant dynamics by the compensation.

Example

It is required to determine suitable closed-loop configurations for the plant below such that the system is optimal with respect to the cost function (7.5).

$$P(s) = \frac{1}{s(s + 1)}$$

Using equation (7.14), we have:

$$\begin{aligned} Y(s)Y(-s) &= 1 + \lambda^2 s(s + 1)(-s)(-s + 1) \\ &= 1 - \lambda^2 s^2 + \lambda^2 s^4 \\ &\triangleq (1 + as + bs^2)(1 - as + bs^2) \end{aligned}$$

Equating coefficients results in:

$$b = \lambda \; ; \; a = \sqrt{2\lambda + \lambda^2}$$

We note that $Y(0) = 1$ so that, using equation (7.18) to optimise with respect to a step input, the optimum transfer function is given by:

$$T_o(s) = \frac{1}{1 + s\sqrt{2\lambda + \lambda^2} + \lambda s^2}$$

We can see the effect on the dynamic response of this system of the weighting to be associated with the plant drive. Table (7.1) lists the changes in resonant frequency w_n and damping ration ζ as the weighting is varied. As the latter is increased the system naturally becomes slower and more damped. Given the plant and the optimum system function, two possible feedback configurations are illustrated in figure (7.1).

TABLE (7.1): Effect of variation of the plant-drive weighting factor.

λ	ω_n	ζ
0.5	1.414	0.791
1.0	1.000	0.866
2.0	0.707	1.000

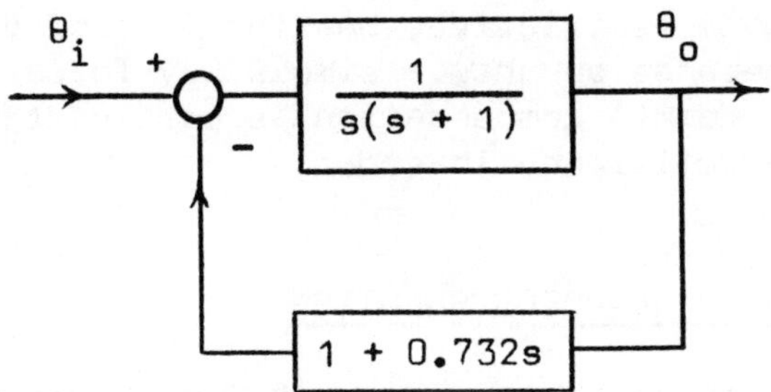

(a) Feedback compensation configuration.

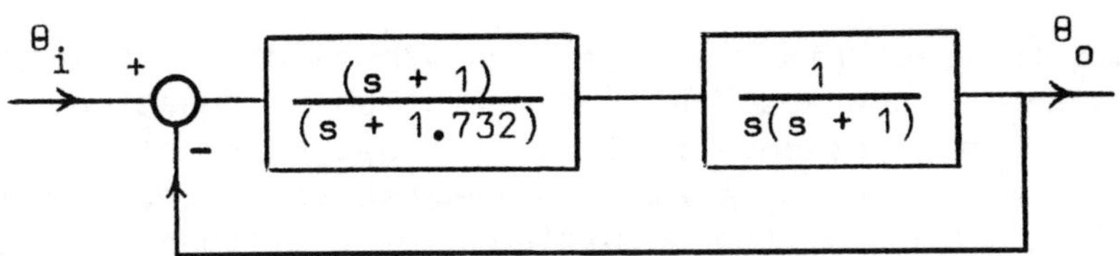

(b) Series compensation configuration.

Fig (7.1): Two possible feedback configurations for the plant of the example, optimised with respect to a step input and with a weighting factor of unity.

It should be noted that had the plant pole of the previous example been located in the RHP, indicative of an unstable open-loop plant, instead of at s = -1, the optimum system function would still be realizable with feedback configuration but would be impossible with the series compensation since this would necessitate the 'cancellation' of a RHP pole. For one thing, exact cancellation can never be relied upon and for another, the unstable mode exists in the homogeneous differential equation quite independently of the modification of the driving function.

The correspondence of the orders of $T_0(s)$ and $P(s)$ is also seen to hold in the example. Should this be considered unduly restrictive, bearing in mind possibly the type of response sought, a simple way forward would be to include 'dummy' compensation as part of the plant to make up any deficiency in order.

7.5. THE ROOT-SQUARE-LOCUS SOLUTION

One untidy aspect of the design method presented in Section (7.4) is the need to quantify the value to be attributed to the plant-drive weighting factor without the benefit of any guidelines. It would be desirable to constrain the plant drive whilst simultaneously having some control over such factors as the system rise time or bandwidth instead of iterating over a range of weighting-factor values as was done with the previous example.

Chang foresaw this need (83) and proposed a root-locus solution to the spectral factorization of equation (7.14). This not only replaced the selection of the weighting factor by that of pole-zero location, and so brought into play the intuition of the designer, but also simplified the solution for higher-order systems for which the process of spectral factorization becomes quite complex. (For computational aspects of spectral factorization see reference 88).

To derive the root-square locus algorithm, let the plant be expressed by equation (7.20).

$$P(s) = \frac{K \prod_{j=1}^{m} (s - z_j)}{\prod_{i=1}^{n} (s - p_i)} \qquad 7.20$$

Equation (7.14) is now written:

$$Y(s)Y(-s) = 1 + \frac{\lambda^2}{K^2} \frac{\prod_{i=1}^{n} (s - p_i)(-s - p_i)}{\prod_{j=1}^{m} (s - z_j)(-s - z_j)}$$

$$= 1 + \frac{\lambda^2}{K^2} \frac{\prod_{i=1}^{n} (p_i^2 - s^2)}{\prod_{j=1}^{m} (z_j^2 - s^2)}$$

We will make the following substitutions:

$$-p_i^2 = P_i \quad : \quad -z_j^2 = Z_j \quad : \quad -s^2 = \Omega$$

We now have:

$$Y(s)Y(-s) = \frac{(\lambda^2/K^2) \prod_{i=1}^{n} (\Omega - P_i) + \prod_{j=1}^{m} (\Omega - Z_j)}{\prod_{j=1}^{m} (\Omega - Z_j)}$$

.... 7.21

It will be noticed that the roots of the numerator polynomial are the poles of the optimum system function which can be found by root-locus methods when the numerator is put into the form of equation (7.22).

$$\frac{K^2}{\lambda^2}\frac{\prod_{j=1}^{m}(\Omega - Z_j)}{\prod_{i=1}^{n}(\Omega - P_i)} = -1 \qquad 7.22$$

The root loci are drawn on the Ω-plane, hence the name applied as the root-square locus method. Suppose we select a set of poles $Q_i = -q_i^2$ lying on the root-loci of equation (7.22). The basis for such a selection will be considered later. However such a selection is, for a given K, tantamount to deciding on the weighting factor λ, since the R^2-locus gain is $(K/\lambda)^2$.

We therefore have:

$$Y(s)Y(-s) = \frac{\prod_{i=1}^{n}(\Omega - Q_i)}{\prod_{j=1}^{m}(\Omega - Z_j)} \qquad 7.23$$

$$Y^2(0) = \frac{\prod_{i=1}^{n}(-Q_i)}{\prod_{j=1}^{m}(-Z_j)} \qquad 7.24$$

Hence,

$$Y(s)Y(-s) = \frac{Y^2(0)\prod_{i=1}^{n}(1 - \frac{\Omega}{Q_i})}{\prod_{j=1}^{m}(1 - \frac{\Omega}{Z_j})} \qquad 7.25$$

We now revert to singularities in the s-plane by the substitution of $-s^2$ for Ω, followed by the taking of the square root of the singularities. There is one set of these for each of Y(s) and Y(-s) of which we retain only those lying in the LHP.

$$Y(s) = \frac{Y(0)\prod_{i=1}^{n}\left(1 - \frac{s}{q_i}\right)}{\prod_{j=1}^{m}\left(1 - \frac{s}{z_j}\right)} \qquad 7.26$$

We therefore have, for the step-input case:

$$T_o(s) = \frac{1}{Y^2(0)} \frac{\prod_{j=1}^{m}\left(1 - \frac{s}{z_j}\right)}{\prod_{i=1}^{n}\left(1 - \frac{s}{q_i}\right)} \qquad 7.27$$

7.5.1. Selection of the Q_i.

It has been emphasised earlier that the correlation between pole-zero locations and system behaviour is not at all straightforward while also insinuating that the intuition of the designer would motivate the selection of the root-square locus poles and zeros. These statements are not incompatible if intuition is founded on a background of familiarity with standard form models of low order. Thus an awareness of such models and their behaviour in relation to their pole-zero geometry will provide the designer with adequate guidelines as to the choice of the Q_i.

For this purpose, Chang relies on either the rise time or the bandwidth to describe the system. (Either will do since both are intimately related). He is less concerned with peak overshoot since, for this type of design, it has a relatively low value of usually less than 20%. In an approximate sense, the bandwidth and the 95% step rise time are proportional and inversely proportional respectively to the distance of the nearest pole to the origin. Thus he restricts the nearest pole to the origin to a distance somewhat less than the required bandwidth or, equivalently, to a distance of π/T_r, where T_r is the time to reach 95% of the steady-state response to a step input.

Thus, the restrictions on distance from the origin in the Ω-plane, bearing in mind its relationship to the s-plane, are:

(a) Rise-Time Specification:

$$\text{Distance} = \left(\frac{\pi}{T_r}\right)^2 \tag{7.28}$$

(b) Bandwidth Specification:

$$\text{Distance} = \omega_B^2 \tag{7.29}$$

<u>Example</u>

Given the plant below, deduce the optimal transfer function of the closed-loop system such that the rise time (95%) is approximately 1.0 second.

$$P(s) = \frac{(s + 3)}{s(s + 1)}$$

The plant singularities are:

$$p_1 = 0 \; ; \; p_2 = -1 \; ; \; z_1 = -3$$

Hence,

$$P_1 = 0 \; ; \; P_2 = -1 \; ; \; Z_1 = -9$$

Further, the distance of the nearest pole in the Ω-plane is to be approximately $(\pi/T_r)^2 = 9.87$. The root-loci are as illustrated in figure (7.2) on which the boxed-in pole represents the desired closed-loop pole lying a distance 9.87 from the origin. By standard root-locus methods, it is found that the break-away points on the real axis lie at -0.46 and -19.54 and that the desired poles lie at:

$$Q_i = -5.86 \pm j7.942$$

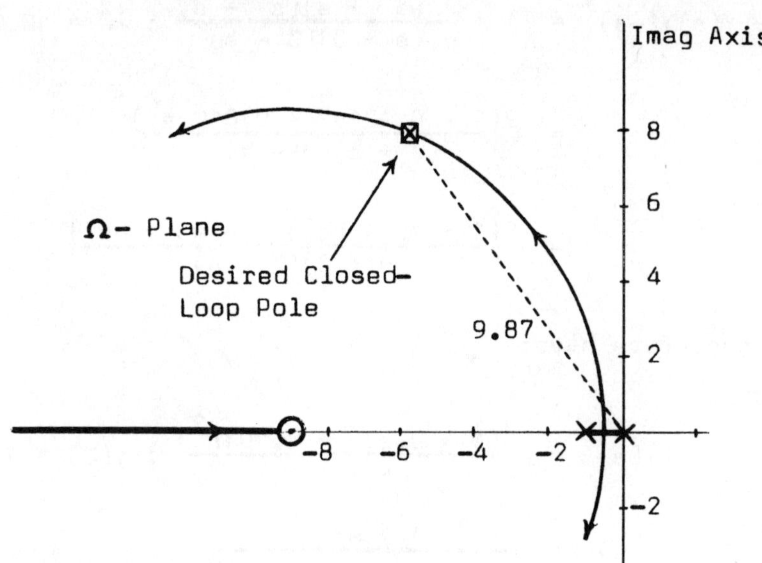

Fig (7.2): Root-square locus plot for the optimum system function with a given rise-time specification.

On taking the square root:

$$q_i = \pm 2.804 \pm j1.416$$

Now, using the magnitude criterion:

$$\text{Gain} = K^2/\lambda^2 = \frac{9.87 \times 9.34}{8.54} = 10.79$$

Thus, if K is left as unity, this value of gain is equivalent to a weighting factor of 0.3. Finally:

$$Y(s) = \frac{s^2 + 5.608s + 9.87}{\sqrt{10.79}(s+3)} \quad : \quad Y(0) = 1$$

$$T_o(s) = \frac{1 + s/3}{1 + 0.568s + 0.101s^2}$$

If we accept the above value for the weighting factor, the design can be verified by using the earlier technique of spectral factorization, thus:

$$Y(s)Y(-s) = 1 + \frac{\lambda^2(s^2 + s)(s^2 - s)}{(s + 3)(3 - s)} : \lambda^2 = 10.79$$

$$= \frac{9(1 - 0.121s^2 + 0.0103s^4)}{(s + 3)(3 - s)}$$

$$= \frac{9(1 + as + bs^2)(1 - as + bs^2)}{(s + 3)(3 - s)}$$

We therefore have:

$$Y(s) = \frac{3(1 + 0.57s + 0.101s^2)}{s + 3} : y(0) = 1$$

$$T_o(s) = \frac{1 + s/3}{1 + 0.568s + 0.101s^2}$$

7.5.2. The Optimal Solution at High Gain

It is of interest to examine the nature of the optimum transfer function as the loop gain increases indefinitely. We are aware from the behaviour of root-loci that those poles in excess of the zeros will, with increasing gain, migrate towards infinity and eventually coincide with asymptotes spread symmetrically with respect to the real axis such that the angle they bear times the pole-zero excess is always an odd multiple of 180°.

Thus, in the Ω-plane, the roots at high gain (ignoring those that have approached within infinitesimal distances of the zeros and therefore have formed dipoles) are given by equation (7.30).

$$Q_k = re^{j(\pi + \frac{2k\pi}{n})} : k = 0,..,n-1 \qquad 7.30(a)$$

... when n is the pole-zero excess and assumes odd values, or:

$$Q_k = re^{j(\frac{\pi}{n} + \frac{2k\pi}{n})} \quad : k = 0,..,n-1 \qquad 7.30(b)$$

... when n assumes even values.

The s-plane roots are found by taking the square root:

$$s = Q_k^{\frac{1}{2}} = \pm r^{\frac{1}{2}} e^{j(\frac{\pi}{2} + \frac{k\pi}{n})} \quad : k = 0,..,n-1 \qquad 7.31(a)$$

... for n odd, or

$$= Q_k^{\frac{1}{2}} = \pm r^{\frac{1}{2}} e^{j(\frac{\pi}{2n} + \frac{k\pi}{n})} \quad : k = 0,..,n-1 \qquad 7.31(b)$$

... for n even.

Thus, for the case of a pole-zero excess of 4, the s-plane roots are:

$$s = \pm r^{\frac{1}{2}}.e^{j(1 + 2k)\pi/8} \qquad 7.32$$

All 2n roots are plotted in figure (7.3) where it is seen that they are equally distributed around a circle of radius $r^{\frac{1}{2}}$ and symmetrical about the real axis. When the n LHP are retained to constitute Y(s) these latter properties will persist and be descriptive of the poles of the optimum transfer function. (Recall that this is the reduced function of order equal to the pole-zero excess which, at high loop gain, is unaffected by the remaining dipoles). A function with this arrangement of poles is known as a Butterworth function and has a frequency response amplitude ratio which has no resonances within the pass-band, hence its alternative name of a 'maximally flat function'.

For all orders, the transient response of such a system

has low percentage overshoot and is accompanied by very heavily damped oscillations. Several such responses are illustrated in figure (7.4).

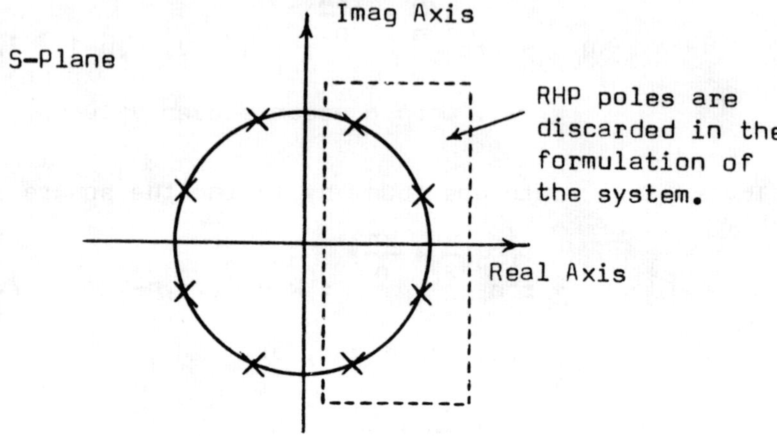

Fig (7.3): Configuration of the excess poles of an optimum system with high gain. (Order = 4).

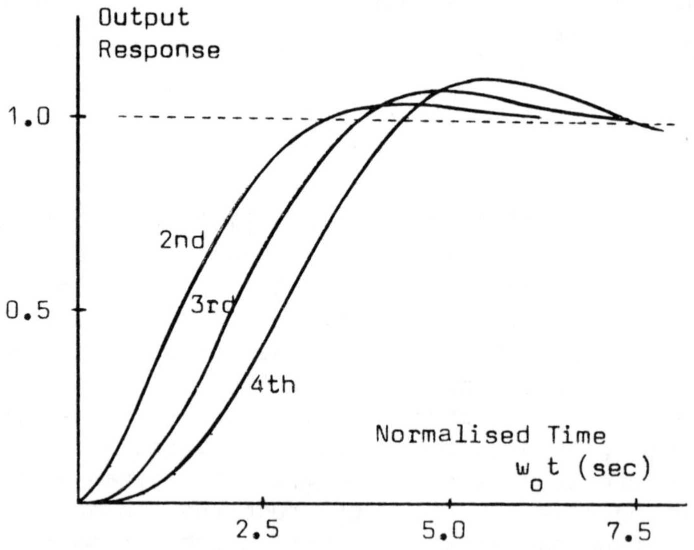

Fig (7.4): Step function responses of normalised Butterworth functions of second to fourth orders.

7.6. AN OBSERVER-FEEDBACK CONFIGURATION

If this chapter had concentrated on the development of an optimal control policy centred on state-variable representation as reviewed in Appendix (E), the most natural way of implementing the control would have been to feed back a linear combination of all the state variables. That is, there would be no compensatory dynamics in the feedback. (Note that this implies all state variables must be available for feedback, or if not, that some or all of them will be estimated using Observers (89) or Optimal Estimators (46,47)). The assumption is of course that transducers are available and can be installed on the hardware and further, since they appear in the equations simply as multipliers, that their dynamics are of such high-frequency capability that they do not interfere with the dynamics of the loop.

However, when optimal control is approached from the point of view of the optimum transfer function, it is apparent that there are many possible feedback configurations which will synthesize the optimal system. For example, unity output feedback with forward-path compensators or combinations of forward-path and feedback path compensation. The fewer the state variables which are fed back however the greater will be the order of the alternative compensation of up to n-1 where n is the order of the plant. This should not weigh too heavily since the cost of compensation is much less than that for the installation of transducers.

If no consideration is of importance in the design other than the achievement of optimality then it is immaterial how the optimum transfer function is synthesized, provided a realizable and stable system is obtained. If however the plant is open-loop unstable then difficulties may be encountered since the optimum design involves cancellation. To overcome this, Chen (84) proposed the use of the Observer configuration of figure (7.5) which has the advantage that, because of the additional compensation introduced by the feedback of the plant input, the poles of the compensators and the cancelled poles can be chosen arbitrarily.

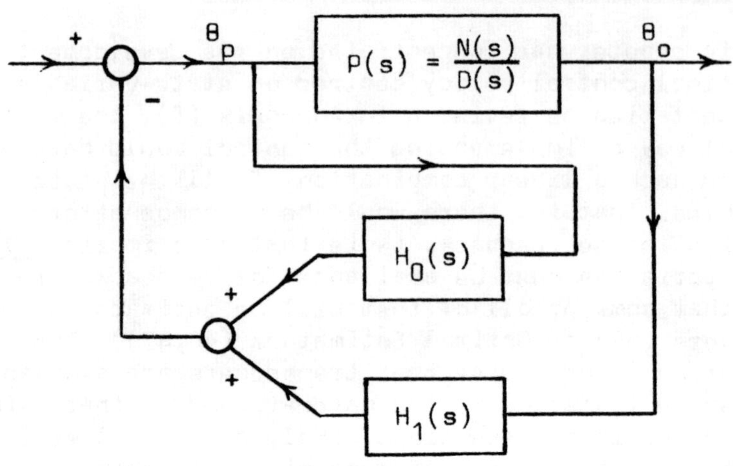

Fig (7.5): Observer feedback configuration for implementation of the optimum transfer function.

Let the plant and feedback be defined by:

$$P(s) = \frac{N(s)}{D(s)} = \frac{b_0 + b_1 s + \cdots + b_n s^n}{a_0 + a_1 s + \cdots + a_n s^n} \qquad 7.33$$

$$H_0(s) = \frac{N_0(s)}{D_c(s)} \qquad 7.34$$

$$H_1(s) = \frac{N_1(s)}{D_c(s)} \qquad 7.35$$

where, $N_0(s) = h_{00} + h_{01}s + \cdots + h_{0(n-1)}s^{n-1}$

$N_1(s) = h_{10} + h_{11}s + \cdots + h_{1(n-1)}s^{n-1}$

Thus if:

$$D_c(s)\{D_f(s)-D(s)\} = d_0 + d_1 s + \cdots\cdot + d_{2n-1}s^{2n-1}$$

where D_f is the required characteristic polynomial.

.. then the coefficients d_i may be computed using equation (7.36):

$$\begin{bmatrix} a_0 & 0 & 0 & b_0 & 0 \ldots & 0 \\ a_1 & a_0 & 0 & b_1 & b_0 \ldots & 0 \\ \ldots\ldots & & & \ldots\ldots & & \\ a_n & a_{n-1} & \ldots & b_n & b_{n-1} & \ldots \\ 0 & a_n & \ldots & 0 & b_n & \ldots \\ \ldots\ldots & & & \ldots\ldots & & \\ 0 & \ldots\ldots & a_n & 0 & \ldots\ldots & b_n \end{bmatrix} \cdot \begin{bmatrix} h_{00} \\ h_{01} \\ \ldots \\ h_{10} \\ h_{11} \\ \ldots \\ h_{1(n-1)} \end{bmatrix} = \begin{bmatrix} d_0 \\ d_1 \\ \cdot \\ \cdot \\ \cdot \\ \cdot \\ d_{2n-1} \end{bmatrix} \quad \ldots .\ 7.36$$

Similar algorithms have been presented by Chen when it is required to feedback some or all of the system's state variables.

7.7. BODE DIAGRAM SYNTHESIS

Time domain synthesis of the optimal control law concludes with the formulation of a matrix equation known as the Riccati Equation (see Appendix E, equation E20) which must be solved to provide the matrix of gain constants defining the state-variable feedback. It was shown however by Leake (85) that this could be circumvented to obtain an approximation to the optimum regulator using only a Bode diagram representation. His derivation was based on the fundamental frequency-domain condition of optimality derived by Kalman (86) known as the Kalman equation (47, pp 280-4, pp 326-33). On simplification the condition becomes:

$$\left|1 + P(s)H(s)\right|^2 = 1 + \left|P(s)\right|^2 \cdot \frac{1}{\lambda} \qquad 7.37$$

where P(s) is the plant function and H(s) is the feedback function equivalent to complete state feedback.

The implication of equation (7.37) is that the LHS is always greater than unity and that the loop function must therefore lie outside the unit circle in the s-plane centred on the Nyquist point. Recall that this is the condition to be satisfied for the sensitivity function to be less than unity at all frequencies. But note also that complete state-variable feedback is equivalent to the feedback of output derivatives up to the order n−1, where n is the order of the plant, which can have serious consequences for noise transmission, especially so for the plant input when L(s)/P(s) can theoretically approach infinity at high frequencies.

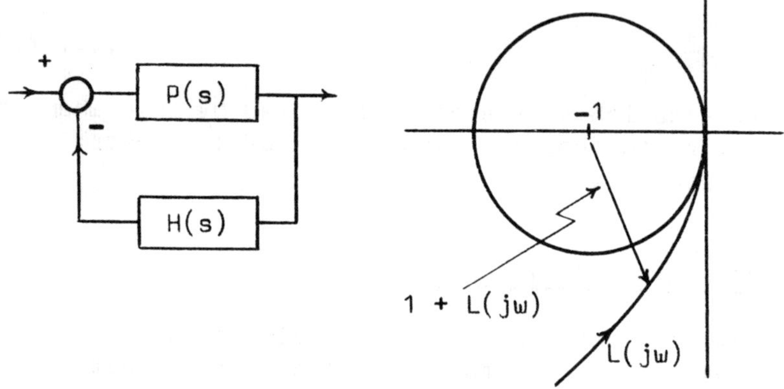

Fig (7.6): Loop function property of an optimal system with complete state feedback.

Leake identified three important properties of equation (7.37):

(a) At low frequencies.

$$|1 + P(s)H(s)| \cong |P(s)| \cdot \frac{1}{\sqrt{\lambda}} \qquad 7.38$$

(b) At high frequencies.

$$|1 + P(s)H(s)| \cong 1 \tag{7.39}$$

(c) At the modified cross-over frequency.

$$|P(s)| = \sqrt{\lambda}$$

$$|1 + P(s)H(s)| = \sqrt{2} \tag{7.40}$$

Conditions (a) and (b) are met by most real systems and definitely so if the plant has at least one integration and a pole-zero excess of one or more.

The steps of the design procedure are as follows:

(1) Sketch the asymptotic Bode diagram of $P(s)/\sqrt{\lambda}$ up to the frequency when the magnitude has fallen to 0db.

(2) Continue the Bode diagram along the 0db-axis. Conditions (a) and (b) are now satisfied.

(3) The junction between the asymptotes of (1) and (2) must now be modified to satisfy condition (c). This is done by multiplying P(s) by a function whose value at the junction frequency w_c is $\sqrt{2}$ and whose zeros will become the poles of the system function.

A convenient function for use in step (3) is the nth-order Butterworth function having an order equal to the pole-zero excess of P(s) function which lies above the 0db axis. Butterworth polynomials are defined:

$$1 + a_1\frac{s}{w_c} + \dots + a_{n-1}\left(\frac{s}{w_c}\right)^{n-1} + \left(\frac{s}{w_c}\right)^{n} \tag{7.41}$$

$$\text{where,} \quad a_{n-k} = \frac{a_{n-k+1}\cos\left\{\frac{\pi(k-1)}{2n}\right\}}{\sin\left(\frac{k\pi}{2_n}\right)}$$

Example

Given the plant:

$$P(s) = \frac{100}{s(1 + s)(1 + 0.2s)}$$

Derive the optimum system function for a plant-drive weighting factor of unity.

Following steps (1) and (2), the Bode diagram of figure (7.7) is drawn.

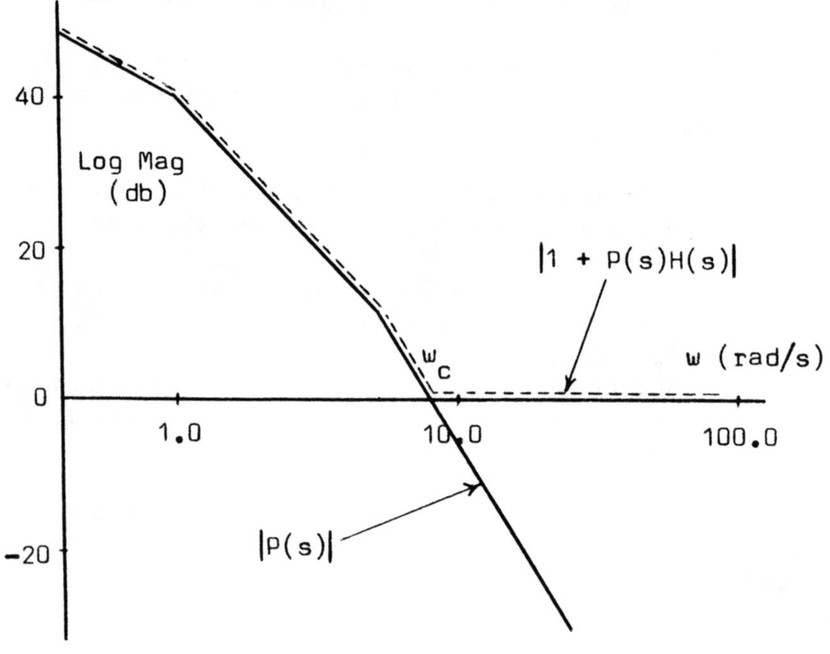

Fig (7.7): Asymptotic Bode diagram construction of the optimum system function prior to modification at w_c with a Butterworth polynomial.

We notice that all of the breaks in the magnitude of $P(jw)$ occur above the 0db axis which therefore calls for a third-order Butterworth polynomial which we obtain from equation (7.41) or Table (7.2). The modified plant function is now formed:

TABLE (7.2): The first three Butterworth Polynomials with a break frequency at w_c.

Order	Polynomial
1	$1 + \frac{s}{w_c}$
2	$1 + \sqrt{2}\,\frac{s}{w_c} + (\frac{s}{w_c})^2$
3	$1 + 2\,\frac{s}{w_c} + 2(\frac{s}{w_c})^2 + (\frac{s}{w_c})^3$

$$\underline{P}(s)B(s) = \frac{500}{s(s+1)(s+5)} \cdot \frac{(s^3 + 2w_c s^2 + 2w_c^2 s + w_c^3)}{w_c^3}$$

where $\underline{P}(s)$ represents the plant function with asymptotic breaks above the 0db axis and $B(s)$ the relevant Butterworth polynomial.

The function $\underline{P}(s)B(s)$ is seen to satisfy conditions (a–c) for the three frequency ranges. Thus, the third-order $B(s)$ function ensures that the magnitude is 0db at high frequencies, is $\sqrt{2}$ at cross-over and approximates to that of the plant at low frequencies.

Now we know that $\underline{P}(s)B(s)$ must be equated to the return difference equation $1 + L(s)$ which is itself, on the assumption of phase-variable feedback of second order, given by:

$$1 + L(s) = 1 + \frac{500K(1 + as + bs^2)}{s(s+1)(s+5)}$$

Hence on comparing coefficients with $\underline{P}(s)B(s)$ we have:

$$K = 1 \; ; \; w_c = 7.937 \; ; \; a = 0.242 \; ; \; b = 0.0197$$

The feedback function is therefore:

$$H(s) = 1 + 0.242s + 0.0197s^2$$

.. and the optimum system function is:

$$T_o(s) = \frac{1}{1 + 0.252s + 0.0317s^2 + 0.002s^3}$$

This method is a convenient and rapid means of determining the optimum function provided that the loop gain is sufficiently high to ensure that all plant singularities occur below the junction frequency w_c. Should this not be the case, the solution will deviate more and more from the true optimal design as the gain is reduced. Conversely, with increasing loop gain, the junction will have a decreasing effect on the solution until, at high gain levels, the true solution for T_o is simply the inverse of B(s). That is, the optimal high-gain solution is a Butterworth function of appropriate order as pointed out previously by Chang.

7.8. SUMMARY

It has been seen that the objective of optimal design, rather than being a search for the 'superlative' system, is the more pragmatic formulation of a closed-loop system according to a particular design approach which involves both the search for that system which minimises a certain performance index and its implementation via a variety of compensators.

The performance index in common use is of quadratic form, chosen primarily for mathematical convenience but nevertheless one which has practical value in that it tends to penalise excessive deviation and to mitigate against plant saturation. Only in a few situations can its associated weighting factor be assigned a specific value, as with fuel or other financial costs. Usually its selection is dependent upon trial-and-error where the final choice is often decided by the consequential dynamics of the closed-loop system.

Two approaches are possible, one based on transfer function representation and the other on state-variable formulation. With the latter, the solution is in the form of a control law which implies the availability of all state variables for feedback. With the former approach however, the design results in an optimal transfer function which may be implemented according to the dictates of the designer. Both methods are amenable to a more classical interpretation, the one using root-loci to give an exact answer and the other to the use of Bode diagrams which lead to an approximate Butterworth characteristic polynomial.

Criticisms have been levelled at this design procedure (38,87) as being unduly restrictive and of questionable engineering feasibility for real and uncertain systems. The more serious of these are listed below and left for the consideration of the reader.

(a) No account is taken of the major problems of plant uncertainty or external disturbance.

(b) There is no control over the ultimate transfer function and its time response.

(c) There is an unnecessary avoidance of dynamic compensation which involves costly measurements which introduce further dynamics into the system, or the alternative of state estimation which relies on plant invariance.

(d) Large amounts of phase advance are necessary with the attendant noise problems.

(e) There is no basis upon which to select the plant weighting factor.

As a technique however it does provide the designer with an explicit solution to the problem of constraining the control effort which may be useful as a guideline, especially in applications with adequate instrumentation and where the plant has been accurately modelled and shows little variation.

APPENDIX A

Algorithms For Low-Order Modelling

A1. LEAST-SQUARES COMPLEX CURVE FITTING VIA THE METHOD OF SANATHANAN AND KOERNER (21)

The modified error function of equation (3.20) may be written:

$$E = \sum_{k=1}^{m} |e(\omega_k)D(\omega_k)_L|^2 . W_{kL} \qquad \text{A1}$$

$$\text{where } W_{kL} = 1/|D(\omega_k)_{L-1}|^2$$

Partial differentiation of E with respect to the unknown coefficients and equating to zero results in a set of equations which may be simplified by the introduction of the following intermediate terms:

$$Q_i = \sum_{k=1}^{m} \omega_k^i W_{kL} \qquad \text{A2(a)}$$

$$S_i = \sum_{k=1}^{m} \omega_k^i R_k W_{kL} \qquad \text{A2(b)}$$

$$T_i = \sum_{k=1}^{m} \omega_k^i I_k W_{kL} \qquad \text{A2(c)}$$

$$U_i = \sum_{k=1}^{m} w_k^i (R_k^2 + I_k^2) W_{kL} \qquad A2(d)$$

where R_k and I_k are the real and imaginary parts of the actual frequency response, w_k is frequency and the summations are over the m points of data.

The set of simultaneous equations is derived:

$$(A).(B) = (C)$$

where,

$$(B) = (a_0, a_1, \ldots, b_1, b_2, \ldots)^T$$

$$(C) = (S_0, T_1, S_2, T_3, \ldots, 0, U_2, 0, U_4, \ldots)^T$$

where the superscript T indicates transposition and it is assumed that the model to be fitted to the measured data has the transfer function:

$$G_m(s) = \frac{a_0 + a_1 s + a_2 s^2 + \ldots}{1 + b_1 s + b_2 s^2 + \ldots.}$$

The matrix (A) can be partitioned as follows:

$$(A) = \left[\begin{array}{c|c} Q & F \\ \hline D & U \end{array}\right]$$

where,

$$Q = \underbrace{\begin{bmatrix} Q_0 & 0 & -Q_2 & 0 & Q_4 & 0 & -Q_6 & 0 & \ldots \\ 0 & Q_2 & 0 & -Q_4 & 0 & Q_6 & 0 & -Q_8 & \ldots \\ Q_2 & 0 & -Q_4 & 0 & Q_6 & 0 & \ldots & & \\ 0 & \ldots & & & & & & & \\ \ldots & & & & & & & & \end{bmatrix}}_{M+1} \quad \updownarrow M+1$$

where M is the order of the model numerator and

where the highest order of Q_i is $i = 2M$.

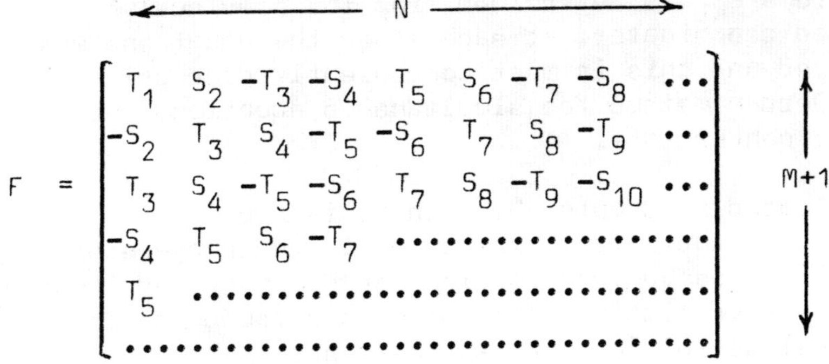

$$F = \begin{bmatrix} T_1 & S_2 & -T_3 & -S_4 & T_5 & S_6 & -T_7 & -S_8 & \cdots \\ -S_2 & T_3 & S_4 & -T_5 & -S_6 & T_7 & S_8 & -T_9 & \cdots \\ T_3 & S_4 & -T_5 & -S_6 & T_7 & S_8 & -T_9 & -S_{10} & \cdots \\ -S_4 & T_5 & S_6 & -T_7 & \cdots \\ T_5 & \cdots \\ \cdots \end{bmatrix} \quad \overset{\longleftarrow N \longrightarrow}{} \quad \updownarrow M+1$$

where N = order of the model denominator and where the highest order of S_i, T_i is $i = N+M$

$$\overset{\longleftarrow \; M+1 \; \longrightarrow}{} \qquad D = \begin{bmatrix} T_1 & -S_2 & -T_3 & S_4 & T_5 & -S_6 & -T_7 & S_8 & \cdots \\ S_2 & T_3 & -S_4 & -T_5 & S_6 & T_7 & -S_8 & -T_9 & \cdots \\ T_3 & -S_4 & -T_5 & S_6 & T_7 & -S_8 & \cdots \\ S_4 & T_5 & -S_6 & -T_7 & \cdots \\ T_5 & -S_6 & -T_7 & \cdots \\ \cdots \end{bmatrix} \quad \updownarrow N$$

$$\overset{\longleftarrow \; N \; \longrightarrow}{} \qquad U = \begin{bmatrix} U_2 & 0 & -U_4 & 0 & U_6 & 0 & -U_8 & 0 & \cdots \\ 0 & U_4 & 0 & -U_6 & 0 & U_8 & 0 & -U_{10} & \cdots \\ U_4 & 0 & -U_6 & 0 & U_8 & 0 & \cdots \\ 0 & U_6 & 0 & -U_8 & \cdots \\ U_6 & 0 & -U_8 & \cdots \\ \cdots \end{bmatrix} \quad \updownarrow N$$

where the highest order of U_i is $i = 2N$

In the first iteration with $W_{kL} = 1$, the solution is exactly as obtained by Levy's method, but in succeeding iterations W_{kL} is determined from the previously computed denominator. At each stage the equations must be solved and this is most conveniently done using the Gauss-Jordan method for simultaneous equations which ensures convergence.

As indicated in Chapter 3, when it is required to obtain a low-order model from experimental frequency-response data, it is convenient to find the dominant model using complex-curve fitting. The author has found the following model useful in a variety of cases:

$$T_m(s) = \frac{1 + a_1 s}{1 + a_1 s + a_2 s^2 + a_3 s^3} \qquad \text{A3}$$

This model is one which, within the dominant frequency range, is representative of a zero velocity-lag system. Should such a model be required, then the following algorithm, derived by constraining the iterative procedure, has been provided by Payne et al (23).

$$\begin{bmatrix} U_2 - 2S_2 + Q_2 & T_3 & S_4 - U_4 \\ T_3 & U_4 & 0 \\ S_4 - U_4 & 0 & U_6 \end{bmatrix} \cdot \begin{bmatrix} a_1 \\ a_2 \\ a_3 \end{bmatrix} = \begin{bmatrix} 0 \\ U_2 - S_2 \\ T_3 \end{bmatrix} \qquad \text{A4}$$

A2. COMPUTATION OF CHEN'S CONTINUED-FRACTION MODEL

Let the actual system and its model be represented by:

$$\text{System:} \quad T(s) = \frac{a_{2,1} + a_{2,2}s + \ldots + a_{2,n}s^{n-1}}{a_{1,1} + a_{1,2}s + \ldots + a_{1,n+1}s^n} \qquad \text{A5}$$

$$\text{Model:} \quad T_m(s) = \frac{r_0 + r_1 s + \ldots\ldots + r_m s^{m-1}}{d_1 + d_2 s + \ldots\ldots + d_{m+1} s^m} \qquad \text{A6}$$

The algorithm for Chen's model is:

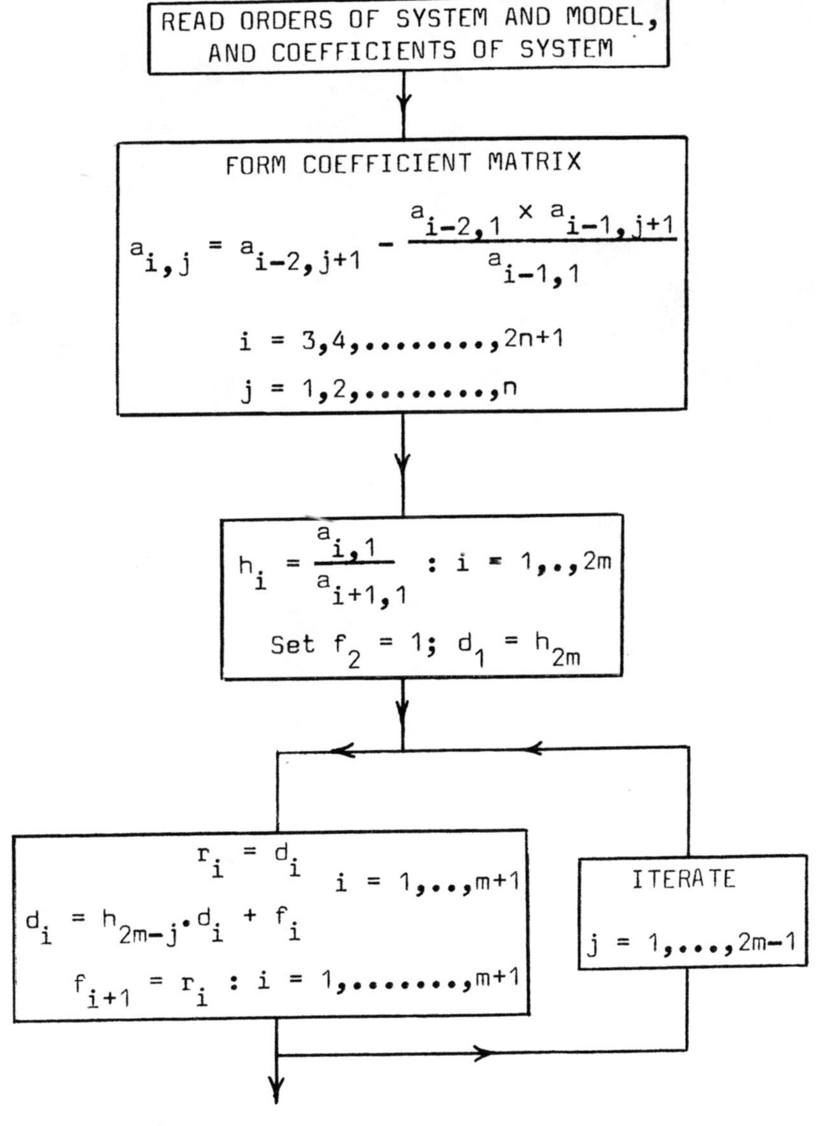

APPENDIX B

Loop-Magnitude Boundary Circles

B1. PARAMETERS OF THE BOUNDARY CIRCLES

A magnitude circle is defined below which is the locus of all points such that the ratio of the magnitudes of the vectors AB/AC (see figure 5.10) is a constant M. Two circles to every P_0/P point will be required, for which:

$$(a) \quad |T| > |T_0| \quad : \quad M = \frac{1}{1 + TT} < 1 \qquad \text{B1(a)}$$

$$(b) \quad |T| < |T_0| \quad : \quad M = \frac{1}{1 - TT} > 1 \qquad \text{B1(b)}$$

where T, T_0 are the actual and nominal system functions and TT is the required tolerance as a fractional quantity.

In the discussion below only the term M will be used where, for the computation of any particular circle, the appropriate value will be used. Note also that the quantity D below is to be treated as a vector about the point +1.
It is seen from figure (B1) that:

$$M = \frac{|1 - x + D + jy|}{|1 - x + jy|} \qquad \text{B2}$$

Equation (B2) now reduces to:

$$(x + \frac{M^2 - D - 1}{1 - M^2})^2 + y^2 = \frac{D^2 M^2}{(1 - M^2)^2} \qquad \text{B3}$$

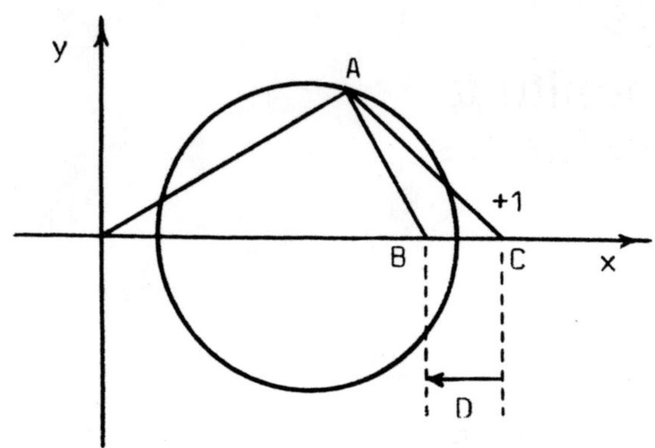

Fig (B1): Definition of the boundary circle without rotation (ie, P_0/P lies on real axis).

Equation (B3) is that of a circle for which:

$$\text{Centre} = (\frac{1 + D - M^2}{1 - M^2}, 0)$$

B4

$$\text{Radius} = \left|\frac{DM}{1 - M^2}\right|$$

Note that the sign of D must be involved to determine the centre of the circle. If now the original set of axes is rotated through an angle α about the point (1,0), we have the arrangement of figure (B2). It can be seen that the following relationships apply:

$$x' = x\cos\alpha + y\sin\alpha + 1 - \cos\alpha$$

$$y' = y\cos\alpha + \sin\alpha - x\sin\alpha \qquad \text{B5}$$

$$|\alpha| \leqslant \pi/2$$

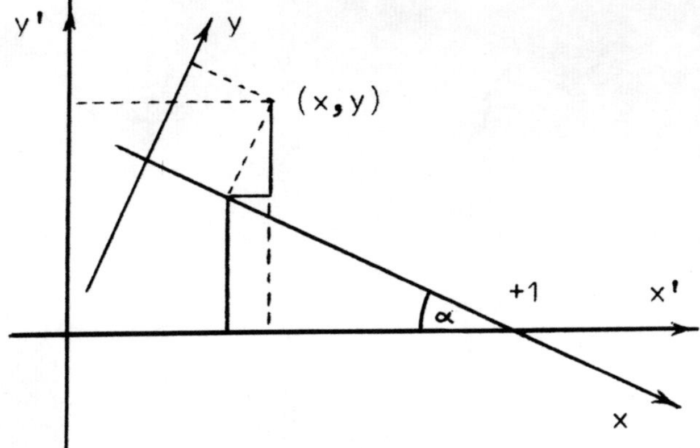

Fig (B2): Relationship between the co-ordinates on rotation of the axes of the boundary circles.

The equations of the circles so rotated are found in the new co-ordinates (x',y') by noting that the centres and radii are given by:

$$\text{Centre} = \left(\frac{1 + D\cos\alpha - M^2}{1 - M^2}, \frac{-D\sin\alpha}{1 - M^2}\right)$$

B6

$$\text{Radii} = \left|\frac{DM}{1 - M^2}\right|$$

where, again, the appropriate M and the correct sign of D is employed.

APPENDIX C
Characteristics of System Functions

The loop function has been seen to be the fundamental characteristic of a feedback control system upon which rests the ability or otherwise of the system to function satisfactorily in service in a disturbing and changing environment despite the designer's ignorance of these influences and the effects of ageing. In view of this it is not surprising that considerable endeavour has been invested over the years by Bode, Horowitz and others in the study of this function, the constraints to be imposed upon it, the costs which must be paid for its benefits and the means for its optimization.

A relevant question to ask is, is there no limit to the benefits which can be achieved? - or, alternatively, are there restrictions which must be imposed on any realizable function? Common sense would indeed suggest the latter. The problem, then, is to establish these constraints.

Bode (31) was the first to apply himself to this question and in doing so developed a number of contour-integral relationships using Cauchy's Theorem which expressed the analytic connection between the real and imaginary components of a system function. Using these, he arrived at what is known as Bode's Ideal Loop Characteristic which, for a long time, stood as the engineer's guideline to the best that could reasonably be achieved in the way of loop-function shaping. His theorems will be outlined in this appendix together with the extensions of Horowitz who succeeded Bode in insisting on the importance of the loop function for any worthwhile design.

We are to be concerned with functions $F(s)$ which are the transforms of the impulse responses of real systems such that:

$$F(j\omega) = A(\omega) + jB(\omega) \qquad \text{C1}$$

where A,B are even and odd functions respectively, $F(s)$ is finite or zero at infinity, has no poles on the imaginary axis and is analytic in the RHP. Further on expansion we have:

$$F(s) = A_0 + B_0's + A_0''s^2 + \ldots \quad \text{: about the origin.}$$

$$F(s) = A_\infty + B_\infty'/s + A_\infty''/s^2 + \ldots \quad \text{: at infinity}$$

THEOREM 1

'The integral of the real component of the function $F(s)$ over all frequencies is determined solely by the value of the imaginary component at infinite frequency'

$$\int_0^\infty \{A(\omega) - A_\infty\}d\omega = \tfrac{1}{2}\pi B' \qquad \text{C2}$$

where superscription denotes the degree of differentiation.

This is the Resistance-Integral Theorem (Bode, pp 280-3) where A_∞ appears in order to allow variation of the integrand according to ω^{-1} to facilitate solution. Bode employed Cauchy's Residue Theorem to derive this particular theorem in order to yield results of importance in the definition of the characteristics of networks but which also, as he showed, led to a result of some significance to feedback systems.

For particular application to feedback systems, the function $F(s)$ may either be the loop or the return-difference function and therefore possibly including poles or zeros which lie on the imaginary axis. This latter property would immediately exclude it as a proper function. However, even with such singularities, an alternative function $\ln G(s)$ is permitted provided $G(s)$ is neither zero nor infinite as s approaches infinity. The latter consideration would disqualify the loop function of any realistic system but allows the use of the return-difference function. Thus, let:

$$F(s) \triangleq \ln\{1 + L(s)\} \qquad \text{C3}$$

We can therefore write:

$$F(s) = \ln|1 + L(j\omega)| + j\phi$$

$$= A(\omega) + jB(\omega) \qquad \text{C4}$$

Now if $L(s)$ approaches zero as s approaches infinity, then:

$$\lim_{x \to 0} \{\ln(1 + x)\} \longrightarrow x$$

$$F(s) \longrightarrow L(s) \; : \text{as } s \longrightarrow \infty$$

Further, it will be seen from the expansion of $F(s)$ about infinity, that if $F(s)$ and $L(s)$ are to approach zero according to s^{-1}, then $A_\infty = 0$ and $B'_\infty \neq 0$. However, since the loop function of any realistic system will decay according to s^{-n}, where $n \geqslant 2$, then B'_∞ will also be zero. We therefore have:

$$\int_0^\infty \ln|1 + L(j\omega)|\, d\omega = 0 \qquad \text{C5}$$

Equation (C5) is of fundamental importance to feedback theory as is observed by noting that the inverse of the return-difference function determines the sensitivity function of the system.

Corollary

> 'In a single-loop feedback system, with a loop function decaying towards zero at infinite frequency at greater than 6 db/octave, the average regerneration or degeneration over the complete frequency spectrum is zero'

This important corollary implies that the areas under the curve of figure (C1), corresponding to the frequency ranges of superior and inferior closed-loop sensitivity as compared to the open-loop situation, are equal, that is that there is as much positive as negative feedback area for all real systems.

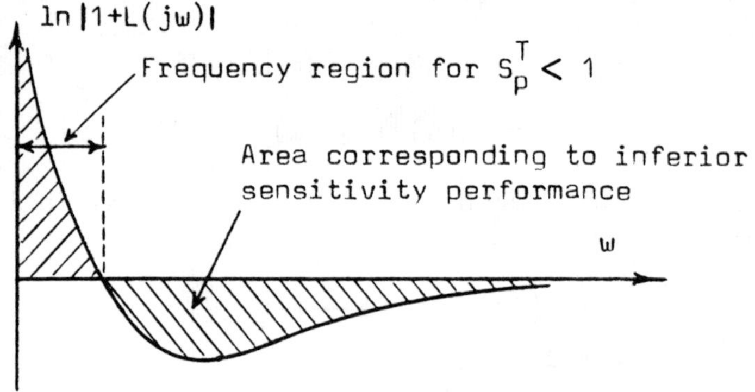

Fig (C1): Variation of $-\ln|S_p^T|$ with frequency showing the equivalence of positive and negative feedback effects.

The engineer will therefore use this characteristic of feedback systems to effect by concentrating the sensitivity reduction area over a comparatively small but useful frequency range with a correspondingly greater loop magnitude.

THEOREM 2

> 'The integral of the imaginary component of a proper function over all frequencies is determined solely by the behaviour of the real component at the extremities of the range'

$$\int_{-\infty}^{\infty} B(w)d(\ln w) = \tfrac{1}{2}\pi(A_{\infty} - A_{0}) \qquad \text{C6}$$

This is the Reactance-Integral Theorem (Bode, pp 286-8) which implies that the integral is independent of the manner in which the real component $A(w)$ varies throughout the frequency range. Thus, for the lead-lag network of figure (C2), where $A_0 = A_{\infty}$, the integral will be zero, indicating equal phase-lead and phase-lag areas.

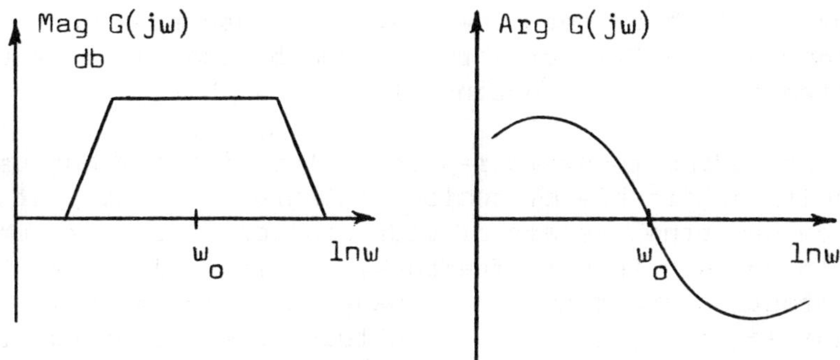

Fig (C2): Magnitude and phase characteristic of a network G(s) with zero difference between the magnitudes at the extreme frequencies, showing equal and opposite phase areas.

Alternatively, from the point of view of feedback, the theorem states that for a given amount of feedback, with $F(s) = \ln L(s)$ and $A(w) = \ln|L(jw)|$, the phase area is constant. Thus, for an unconditionally stable system where the phase at frequencies less than cut-off is to be greater than -180^{0}, the implication of a high degree of feedback may mean the control of $L(jw)$ over a range of frequencies much greater than the required closed-loop bandwidth. On the other hand, if it is required to decrease $|L(jw)|$ as quickly as possible, then $B(w) = \arg L(jw)$ must be kept as large as possible.

Should the loop function have a magnitude which approaches infinity and zero as s approaches zero and infinity respectively, then to apply equation (C6) usefully, the loop function must be suitably modified at sufficiently high and low frequencies such that A_0 and A_∞ are finite without materially affecting the phase in the useful region. We can now rewrite equation (C6) as:

$$\int_{-\infty}^{\infty} B(u)du = \frac{\pi}{40}(A_\infty - A_0) \qquad \text{C7}$$

where $u = \log_{10} w$ and A_0, A_∞ are expressed in decibels.

Thus for 40db of feedback say, the phase area must extend over a frequency range greater than 10:1 if the system is to remain unconditionally stable.

The preceding theorems represent the simplest but useful results obtainable by contour integral analysis but which nevertheless are of wide application. Thus, when one's concern is with feedback, Theorem (2) gives information about the loop-phase characteristic which accompanies a given loop-magnitude cut-off. However, it may be necessary at times to be in possession of more detailed information regarding the relationship between the real and imaginary components.

THEOREM 3

> 'If the real or the imaginary component of F(s) is known over the complete frequency spectrum, this is sufficient to determine F(s) completely'

$$B(w) = \frac{-2w}{\pi}\int_0^\infty \frac{A(x)dx}{(w^2 - x^2)} \qquad \text{C8(a)}$$

$$A(w) - A_\infty = \frac{2}{\pi}\int_0^\infty \frac{xB(x)dx}{(w^2 - x^2)} \qquad \text{C8(b)}$$

The statements of equations (C8a,b), referred to by Horowitz (1963, pp 307-9) as the Real (Imaginary)-Part Sufficiency Theorems, highlight the inter-dependence of the magnitude and phase characteristics such that if either one is specified at any given frequency, there is a corresponding constraint upon the other.

For computational purposes, Bode manipulated equation (C8) into the following form:

$$B(\omega_o) = \frac{1}{\pi}\int_{-\infty}^{\infty} \frac{dA}{du} . \ln \coth\frac{|u|}{2} du \qquad C9(a)$$

$$A(\omega_o) - A_\infty = \frac{-1}{\pi\omega_o}\int_{-\infty}^{\infty} \frac{d(\omega B)}{du} \ln \coth\frac{|u|}{2} du \qquad C9(b)$$

$$\text{where } u \triangleq \ln(\omega/\omega_o)$$

It is seen that the phase characteristic is proportional to the derivative of the magnitude characteristic on a logarithmic scale, weighted by a logarithmic term which places especial emphasis on the rate of change in the vicinity of the frequency $\omega = \omega_o$. One particular case of interest is that for which the magnitude derivative is constant over all frequencies and of value k. Thus:

$$B(\omega) = \frac{k}{\pi}\int_{-\infty}^{\infty} \ln(\coth \frac{|u|}{2}) du = k.\frac{\pi}{2} \text{ rad.} \qquad C10$$

The unit of rate of change can be put into familiar terms by considering the change in $A(\omega) = \ln|L(j\omega)|$ over one decade of frequency. From the definition of u this is obviously ln(10) or, in other words since $A(\omega)$ is expressed in nepers where 1 neper = 8.686db, $L(j\omega)$ has changed in magnitude by 20db. Thus, for every unit of slope of 20 db/decade, the phase characteristic will be constant and of value 90°. At frequencies remote from ω_o, the effect of the weighting can be approximated by noting that:

$$\ln(\coth \frac{|u|}{2}) = \ln\left|\frac{\omega + \omega_o}{\omega - \omega_o}\right| \qquad C11$$

Further approximation leads to:

$$\ln(\coth\frac{|u|}{2}) \cong 2(\omega_0/\omega) \quad : \quad \text{for } \omega \gg \omega_0$$

$$\cong 2(\omega/\omega_0) \quad : \quad \text{for } \omega \ll \omega_0 \qquad \text{C12}$$

In order to facilitate the graphical computation of the relationship between the magnitude and phase, Bode approximated the magnitude curve by a series of straight line segments whose individual contributions to the phase at any frequency were obtained from sets of charts (Bode, pp 337–359).

<u>THEOREM 4</u>

> 'If the real(imaginary) part of a proper function is specified over the frequency range $0 < \omega < \omega_0$, and the imaginary(real) part is specified for $\omega > \omega_0$, then the function is completely specified'

$$\int_0^{\omega_0} \frac{A(x)dx}{(\omega^2-x^2)(\omega_0^2-x^2)^{\frac{1}{2}}} + \int_{\omega_0}^{\infty} \frac{B(x)dx}{(\omega^2-x^2)(x^2-\omega_0^2)^{\frac{1}{2}}}$$

$$= \begin{cases} \dfrac{-0.5\pi B(\omega)}{\omega(\omega_0^2-\omega^2)^{\frac{1}{2}}} & \text{for } \omega < \omega_0 \\[2ex] \dfrac{0.5\pi A(\omega)}{\omega(\omega^2-\omega_0^2)^{\frac{1}{2}}} & \text{for } \omega > \omega_0 \end{cases} \qquad \text{C13(a)}$$

$$\int_0^{\omega_0} \frac{xB(x)dx}{(\omega^2-x^2)(\omega_0^2-x^2)^{\frac{1}{2}}} - \int_{\omega_0}^{\infty} \frac{xA(x)dx}{(\omega^2-x^2)(x^2-\omega_0^2)^{\frac{1}{2}}}$$

$$= \begin{cases} \dfrac{0.5\pi A(\omega)}{(\omega_0^2-\omega^2)^{\frac{1}{2}}} & \text{for } \omega < \omega_0 \\ \dfrac{0.5\pi B(\omega)}{(\omega^2-\omega_0^2)^{\frac{1}{2}}} & \text{for } \omega > \omega_0 \end{cases} \qquad C13(b)$$

Equation (C13a) enables the determination of the real and imaginary components, each over mutually exclusive frequency regions separated by ω_0, when the imaginary and real components respectively are known over the same region. Thus, $A(\omega)$ is assumed known for $\omega < \omega_0$ and $B(\omega)$ for $\omega > \omega_0$ whilst, when the roles are reversed, equation (C13b) is applicable. This theorem is of major importance in the evaluation of optimum characteristics of the cut-off to be attributed to the loop function.

CHARACTERISTICS OF THE IDEAL LOOP-FUNCTION

It has been seen that within the useful frequency range of the system it is desirable to maintain the magnitude of the loop function at sufficiently high levels as will result in satisfactory sensitivity and disturbance rejection performance. This will be, say, in the frequency range $\omega < \omega_0$. Additionally, in order to alleviate difficulties with troublesome plant dynamics at the higher frequencies, to reduce the amount of knowledge concerning the plant which is necessary and to cut the costs of instrumentation for high-frequency operation, <u>the loop magnitude must be reduced as rapidly as possible for $\omega > \omega_0$ without producing an accompanying phase shift greater than some prescribed amount for stability reasons</u>. The need to enforce a rapid cut-off gains priority when the difficult problem of measurement noise is considered (see section 4.4).

However, although there is a need for rapid cut-off, we have seen as a consequence of Theorem (3), that phase shift is broadly proportional to the rate at which the gain changes and which is therefore determined by the degree of stability demanded. For example, the requirement for a phase margin of 30° constrains the maximum rate of cut-off of the loop-magnitude function to only 10 db/octave. To gain an appreciation of what might constitute an ideal cut-off characteristic let us, from a consideration of the previous section, stipulate some constraints to be imposed upon the loop function.

a. From Corollary of Theorem (1) it will be insisted that ω_o should be as low as possible consistent with satisfaction of the dynamic specifications (9).

b. From Theorem (2), the maximum rate of cut-off will be achieved by requiring that argL(jω) should be negative and as large as possible over all frequencies.

c. Theorem (4) reminds us that if magL(jω) is defined for $\omega < \omega_o$, then argL(jω) can be selected independently only for $\omega > \omega_o$.

d. For <u>unconditional</u> <u>stability</u>, argL(jω) must be greater than -180° for $\omega < \omega_c$, where ω_c is the cross-over frequency which, as already noted, places constraints on the cut-off rate according to Theorem (3).

e. As the frequency approaches infinity, argL(jω) will approach $-90n^o$, where n is the pole-zero excess of the loop function. This may well give rise to some conflict in satisfying observation (b) above.

Bearing the above points in mind, a desirable loop-phase characteristic is shown in figure (C3) where the phase margin is maintained for $\omega_o < \omega < \omega_c$, a tolerable rate of increase of phase lag is allowed over $\omega_c < \omega < \omega_1$ to permit a satisfactory gain margin and thereafter argL(jω) is allowed to assume large negative values. Eventually, however, in the region of ω_2, the phase lag must decrease to that value determined by the pole-zero excess.

A major problem in attempting to adopt this approach to synthesis is that, in accordance with Theorem (4),

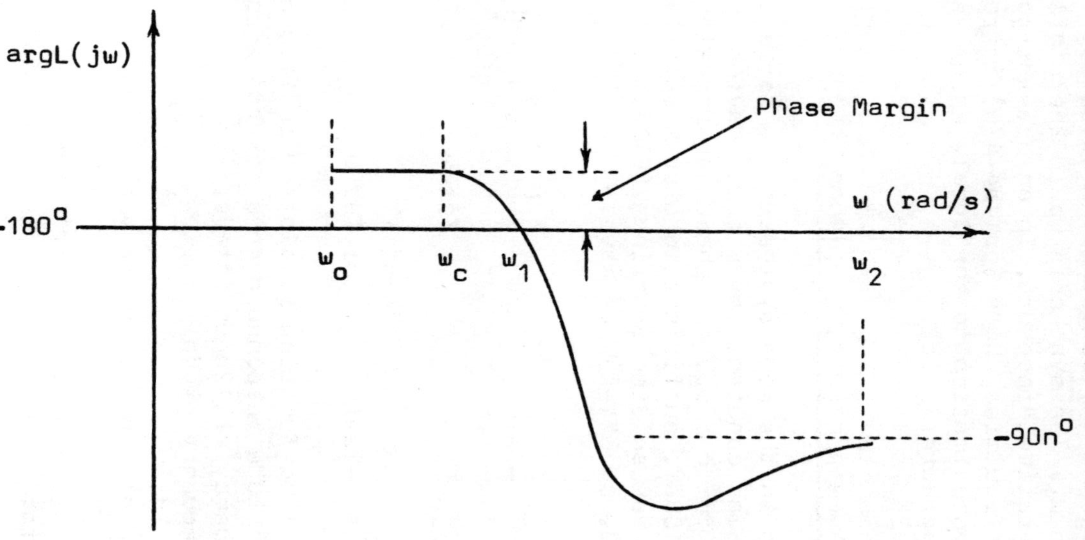

Fig (C3): A desirable phase-lag characteristic for the loop function to satisfy sensitivity and minimum loop-bandwidth.

having specified the phase characteristic we have simultaneously specified the magnitude characteristic for $\omega > \omega_o$ and there is no certainty that it will fulfil our requirements. That is, it may be other than 0 db at ω_c and may be totally unsatisfactory between ω_c and ω_1. The synthesis would therefore be an arduous iterative procedure. In an attempt to overcome these difficulties, Bode (pp 454-476) introduced a practical approximation to the ideal, known as the 'Ideal Bode Characteristic'.

THE IDEAL LOOP-CHARACTERISTIC OF BODE

The first step in the Bode synthesis was to assume that magL($j\omega$) was constant for $\omega < \omega_o$, having a value of unity and, for $\omega > \omega_o$, the phase lag was also constant at $-180a^o$, sufficient to give a satisfactory phase margin. Now solving equations (C13a,b), see Horowitz (1963, pp 321-3), we have:

$$\log_{10}|L(j\omega)| = -2a\log_{10}\{(\omega/\omega_o) + \left[(\omega^2/\omega_o^2) - 1\right]^{\frac{1}{2}}\}$$

$$\ldots \text{ for } \omega > \omega_o \qquad \text{C14(a)}$$

$$\arg L(j\omega) = -2a\sin^{-1}(\omega/\omega_o)\text{: for } \omega < \omega_o \qquad \text{C14(b)}$$

The resulting loop function is plotted in figure (C4) where the magnitude response decays towards zero at infinite frequency at 12a db/octave.
Thus, at a frequency given by:

$$\omega = 2^y\omega_o \quad : \text{ where } y = \text{number of octaves}$$

... the decrease in gain from the low-frequency value of 0db is:

$$\Delta G = 12a(1 + y) \text{ db} \qquad \text{C15}$$

Suppose, for example, that it is required to have a useful frequency-band loop gain of 40db, then:

$$w_c = 2^y w_o \quad \text{and} \quad \Delta G = 40\text{db}$$

If it is required to have a phase margin of 40°, we have that a = 0.777 and hence that y = 3.286. The loop function must therefore be controlled up to a cross-over frequency of $w_c = 9.75w_o$.

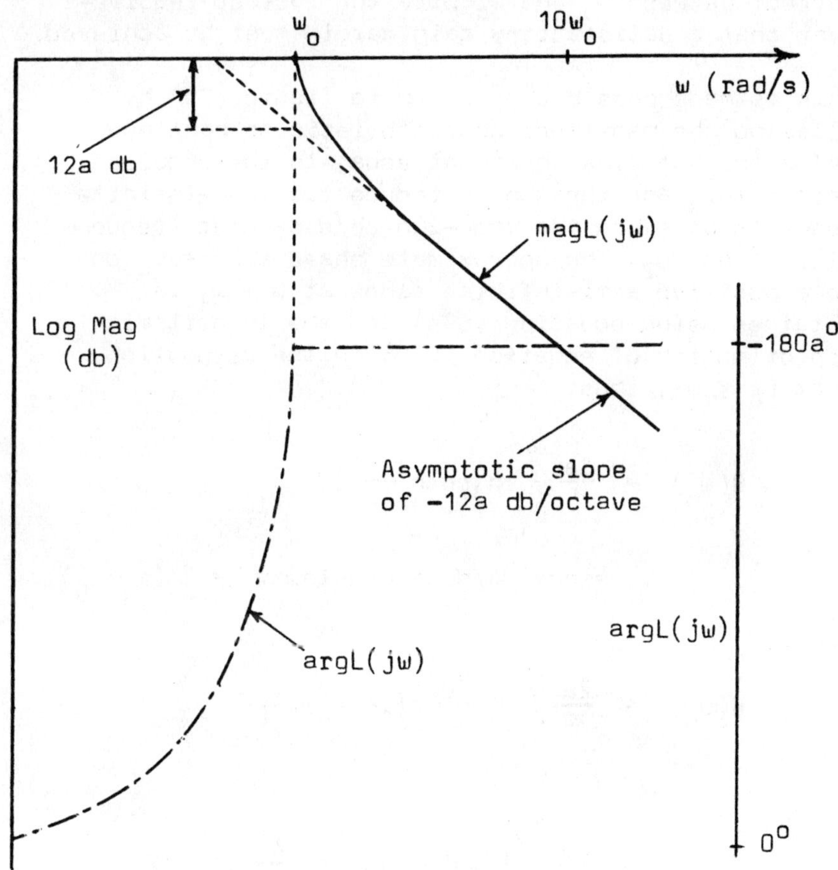

Fig (C4): First stage in the synthesis of the Ideal Bode loop characteristic.

We note however that there is a discrepancy between the high-frequency asymptotic rate of cut-off of 12a db/octave of figure (C4) and the earlier observation (e) requiring a cut-off rate of 6n db/octave. Since, for realistic systems, $n \geqslant 2$, the rate of cut-off at some higher frequency w_2 will have to be increased, which will result in increased phase lag at the lower frequencies according to Theorem (2), with the consequential transgression of the original specifications. Step two of the synthesis thus consists of the junction of the ideal characteristic and the actual high-frequency characteristic such that the phase lag is correct between w_o and w_c with the further requirement that a satisfactory gain margin must be achieved.

This is made possible as shown in figure (C5) by allowing the magnitude characteristic to continue below the 0db line an amount equal to the required gain margin and then to introduce two semi-infinite segments of slope 40a and -20n db/decade at frequencies w_1 and w_2. The approximate phase at $w = w_o$ due to a positive semi-infinite slope at $w = w_1$ is obtained using equation (C9a) and the logarithmic approximation of equation (C12) on the assumption that $w_o \ll w_1$, thus:

$$\theta(w_o) = \frac{2a}{\pi}\int_{u_1}^{\infty} \ln\{\coth\frac{|u|}{2}\}du$$

.. since $dA/du = 0$ below $u_1 = \ln(w_1/w_o)$.

Hence,

$$\theta(w_o) = \frac{2a}{\pi}\int_{u_1}^{\infty} -\frac{2w_o}{w}\, d\{\ln(w_o/w)\}$$

$$= \frac{4a}{\pi}\int_{0}^{w_o/w_1} d(w_o/w) = \frac{4a}{\pi}(w_o/w_1)$$

Similarly, for the negative semi-infinite slope at $w = w_2$:

$$\theta'(w_o) = -\frac{2n}{\pi}(w_o/w_2)$$

Thus, for zero phase effect at w_o, we must have:

$$\frac{n}{a} = 2(\omega_2/\omega_1) \qquad \text{C16}$$

Equation (C16) permits the completion of the ideal characteristic. Note that if $G_1 = 20\log_{10}\text{magL}(j\omega)$ for $\omega < \omega_0$ and G_2 is the required gain margin in decibels, then from equation (C15):

$$y = \frac{G_1 + G_2}{12a} - 1 \; : \; \omega_1 = 2^y \omega_0 \qquad \text{C17}$$

The combination of equations (C16,17) is useful for estimating the range of frequencies over which control of the loop function must be exercised to obtain the benefits of feedback over the range $0 < \omega < \omega_0$. Thus, if 40db of feedback are required with gain and phase margins of 10db and 30°, we find that $\omega_2 = 9.6n\omega_0$. For example, a system with a pole-zero excess of only four must have the loop controlled up to a frequency almost forty times that of its useful range.

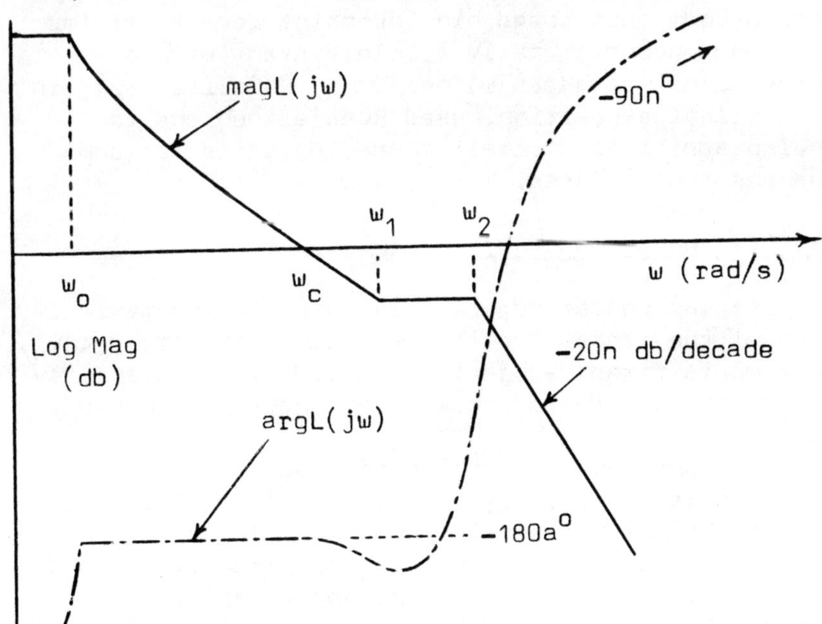

Fig (C5): The ideal Bode characteristic allowing for the true high-frequency characteristic of the loop function.

The loop characteristic just derived is a useful guideline as to the fastest possible cut-off achievable, subject to the definition of the proper function and the nature of the constraints, and although referred to as the 'ideal' function is not, as pointed out by Bode himself (Bode, pp 471-76), the best which can be achieved. We have seen from Theorem (2) that the amount of feedback can be increased by increasing the phase lag. However, an examination of figure (C5) shows that there is scope for improvement in this respect since the phase lag above w_1 can be increased without changing the phase margin. Bode considered a number of modifications to the step junction which, at best, resulted in an increase of feedback of about 4db, when $n = 3$, to 8db when $n = 6$. Whether the improvement is significant and justifies the added complexity is left to the reader's judgement.

ALTERNATIVE 'IDEAL' CHARACTERISTICS

Whereas Bode restricted his investigations to minimum phase and unconditionally stable systems with a defined constant magL(jw) below w_o, Horowitz (32), in an excellent exposition, used Bode's theorems to develop additional 'ideal' characteristics for some less restricted cases.

Maximization of MagL(jw)

Horowitz was concerned, in this case, in the maximization of the area under the magL(jw) curve for $w < w_c$, where w_c is fixed, subject to unconditional stability and to the most rapid decrease of magL(jw) above w_c.

It is obvious that, according to Theorem (2), this objective is achieved by ensuring that argL(jw) assumes the maximum permissible phase lag over the complete frequency spectrum and specifically for $w < w_o$ where argL(jw) > $-180a^o$. The phase variation of figure (C6) was selected. It is seen that the increase in phase lag between w_c and $m_1 w_c$ is controlled to ensure an adequate gain margin, after which it is allowed to increase rapidly to its high-frequency

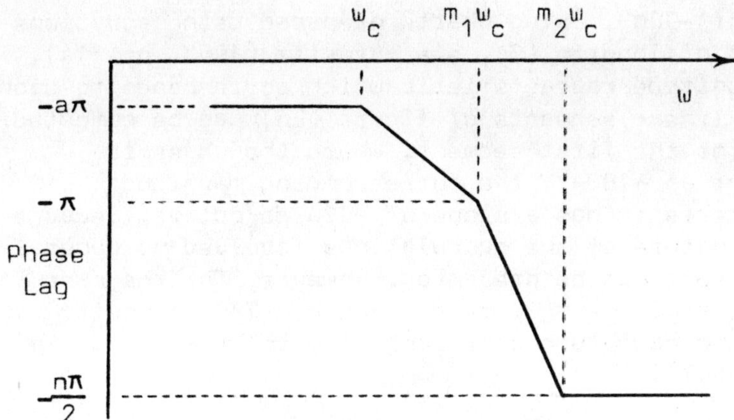

Fig (C6): The assumed phase-lag variation for the maximization of the magnitude of the loop function for $w < w_c$.

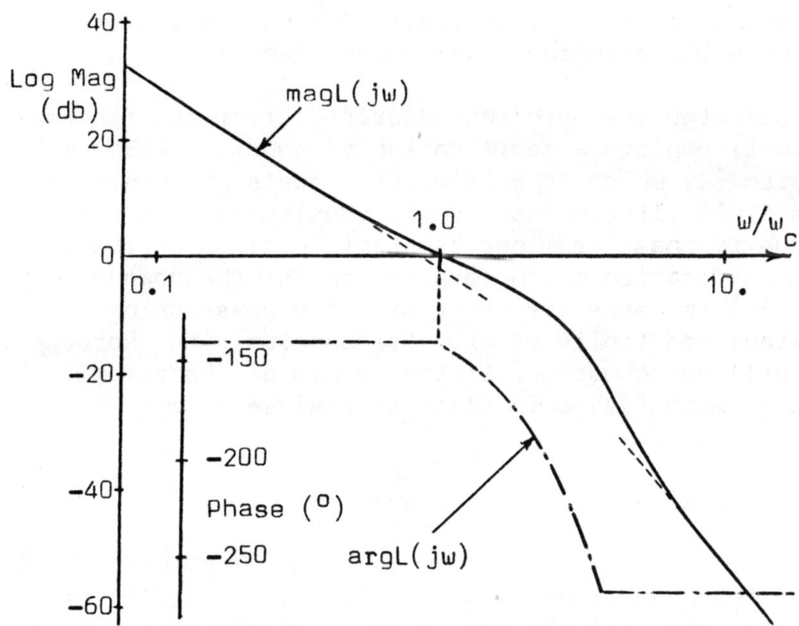

Fig (C7): The 'ideal' characteristic to maximise the loop magnitude below w_c under the constraint of unconditional stability.

value of $-90n^o$. Using charts prepared using equations (C9a,b) of Theorem (3), see Horowitz (1963, pp 314), the magnitude characteristic which corresponds to each of the linear segments of figure (C6) can be computed. Thus, for the first segment, where the phase is constant at $-180a^o$, the corresponding magnitude characteristic has a slope of $-12a$ db/octave. Because of the nature of the calculations involved no generalized result can be presented. However, for the case when $w_c = 10$, $m_1 = 2$, $m_2 = 4$ and $a = 7/9$, Horowitz found the magnitude characteristic to be as shown in figure (C7).

Relaxation of Unconditional-Stability Constraint

If the unconditional-stability constraint of the previous characteristic is relaxed then it will be obvious from figure (C8) that the phase-lag area can be increased considerably with a consequent increase in the permissible feedback. Figure (C8b) illustrates a possible phase characteristic for such a system whilst still ensuring closed-loop stability.

To determine the gain characteristic from the figure, Horowitz employs a modification of equation (C8b) of Theorem (3) which he applies (in Bode's phraseology) to a finite line phase segment consisting of a ramp change in phase from one constant level to another. The modification of the expression for the change of magL(jw) is necessary since with the phase being constant and finite at high frequencies, the expression will not converge. If the method of Theorem (3) is used with F(s)/s in place of F(s) we find:

$$A(w) - A_0 = \frac{2w^2}{\pi}\int_0^{\infty} \frac{Bx^{-1}dx}{w^2 - x^2} \qquad \text{C18}$$

Using the example of figure (C7), equation (C18) is used to construct the amplitude characteristic, first to the major step from $-m\pi$ to $-a\pi$ of figure (C8b) and then, by the introduction of a further finite line phase segment to achieve the required phase margin and a semi-infinite slope (resulting in a step in the

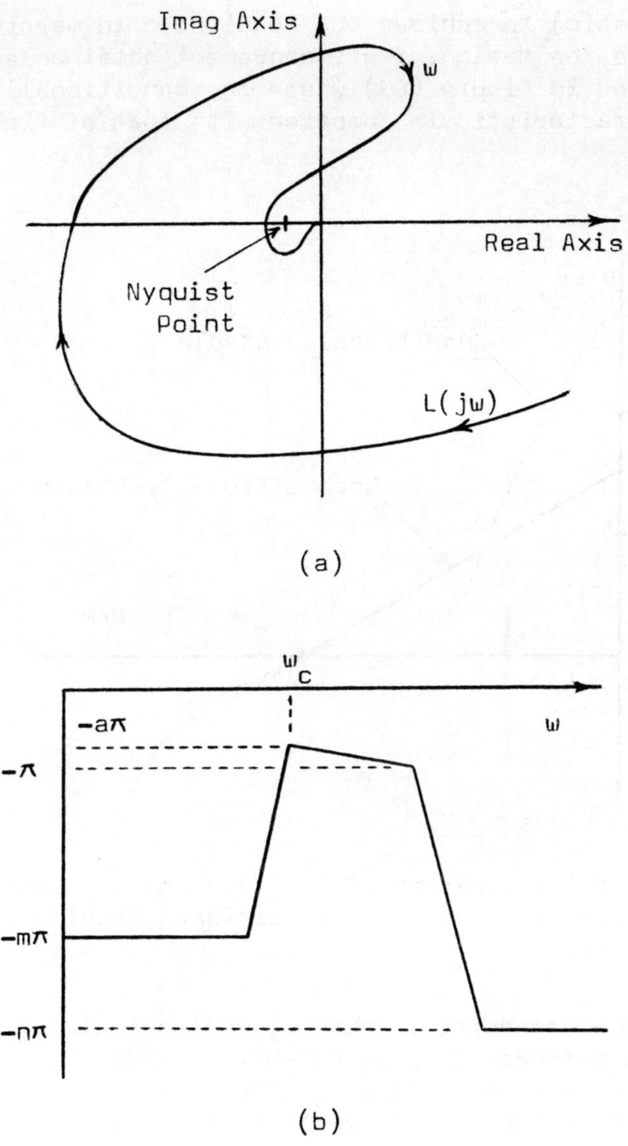

Fig (C8): A possible loop function L(jw) for a conditionally-stable system, (a), with a phase lag characteristic used to determine the corresponding loop-magnitude characteristic, (b).

characteristic) to achieve the required gain margin, to complete the design. The improvement obtainable is high-lighted in figure (C9) where the conditionally-stable characteristic is compared with that of figure (C7).

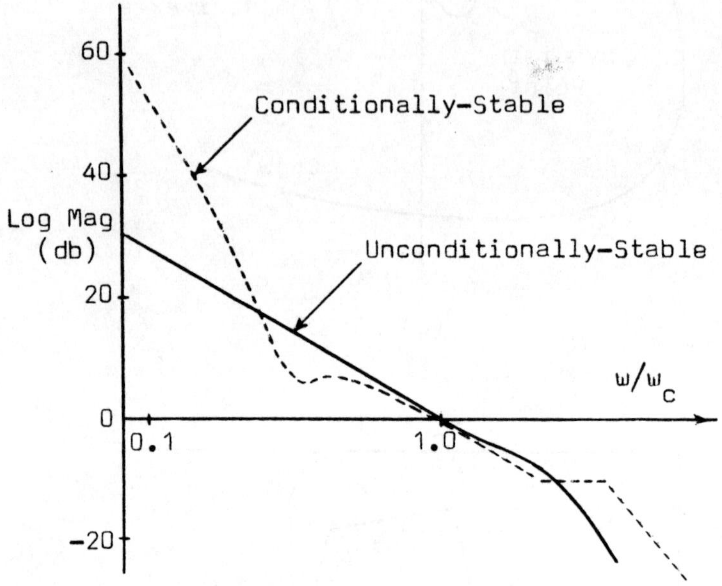

Fig (C9): Illustration of the increase in feed-back which is possible if conditional stability is accepted.

It should be remembered, however, that any element within the feedback loop which suffers from gain variation will have an adverse effect upon the stability of a conditionally-stable system.

APPENDIX D
Parseval's Theorem

D1. PARSEVAL'S THEOREM

Equation (D1) is that of an integral which occurs frequently in systems analysis.

$$I = \int_{-\infty}^{\infty} x(t)y(t)dt \qquad \text{D1}$$

where x,y are Laplace transformable.

The integral may be transformed into frequency domain terms by writing y(t) as the inverse Laplace tranform of Y(s). Thus:

$$I = \frac{1}{2\pi j} \int_{-\infty}^{\infty} \int_{-j\infty}^{j\infty} x(t)Y(s)e^{st} ds.dt \qquad \text{D2}$$

Now on changing the order of integration:

$$I = \frac{1}{2\pi j} \int_{-j\infty}^{j\infty} Y(s) \int_{-\infty}^{\infty} x(t)s^{st} dt.ds \qquad \text{D3}$$

The righthandmost portion of the above integral can be recognized as the Laplace transform of x(t) with the difference that the sign of the variable 's' is positive. It can therefore be written as X(-s) and when substituted into equation (D3) gives expression to Parseval's Theorem as:

$$\int_{-\infty}^{\infty} x(t)y(t)dt = \frac{1}{2\pi j}\int_{-j\infty}^{j\infty} X(-s)Y(s)ds \qquad D4$$

Note that a particularly useful identity is that of equation (D5) relating the integral over all time of the square of the time function x(t).

$$\int_{-\infty}^{\infty} x^2(t)dt = \frac{1}{2\pi j}\int_{-j\infty}^{j\infty} X(s)X(-s)ds \qquad D5$$

D2. TABULATED INTEGRALS

The solution of the right-hand integral of equation (D5) is made particularly simple by the use of the tabulated integrals of Newton, Gould and Kaiser (66). Thus, given:

$$I_n = \frac{1}{2\pi j}\int_{-j\infty}^{j\infty} \frac{c(s)c(-s)}{d(s)d(-s)}\, ds \qquad D6$$

$$\text{where, } c(s) = \sum_{k=0}^{n-1} c_k s^k$$

$$d(s) = \sum_{k=0}^{n} d_k s^k$$

The polynomial d(s) has zeros in the LHP only. The reference to these integrals lists I_n from n = 1 to n = 10, of which only the first three are presented here.

$$I_1 = \frac{c_0^2}{2d_0d_1}$$

$$I_2 = \frac{c_1^2d_0 + c_0^2d_2}{2d_0d_1d_2}$$

$$I_3 = \frac{c_2^2d_0d_1 + (c_1^2 - 2c_0c_2)d_0d_3 + c_0^2d_2d_3}{2d_0d_3(d_1d_2 - d_0d_3)}$$

APPENDIX E

Extremization and Optimality

E1. EXTREMISATION OF FUNCTIONS SUBJECT TO CONSTRAINTS

A function of a number of variables has an extreme value (maximum or minimum) at a point if and only if the partial derivatives of that function with respect to all variables are zero, assuming of course that there are no constraints on the values which may be assigned to those variables. When this is not the case, a mathematical manoeuvre due to Lagrange must be employed known as the method of undetermined multipliers.

Consider a function of say only two variables x and y for simplicity of representation, f(x,y), sketched in figure (E1) as contours in the xy-plane. Obviously an extreme value exists at point A, but if, additionally, the requirement is to find the extreme value of f(x,y) subject to the constraint that g(x,y) = 0, then the extreme value now occurs at point B. It is seen that there will be some contour of f which will be tangential at B to g(x,y) = 0 and which will therefore have a gradient at B colinear with that of the constraint function, but not necessarily of the same magnitude. We can say therefore, for this particular function of two variables, that:

$$\frac{\partial f}{\partial x} = -\lambda\frac{\partial g}{\partial x} \quad \text{and} \quad \frac{\partial f}{\partial y} = -\lambda\frac{\partial g}{\partial y} \qquad \text{E1}$$

The multiplier λ is a scalar relating the two gradients. Now define a new function $f'(x,y)$ such that:

$$f' = f + \lambda g \qquad \text{E2}$$

It will be seen that f' has an extreme value, free of constraint, at point B since:

$$\frac{\partial f'}{\partial x} = \frac{\partial f}{\partial x} + \lambda\frac{\partial g}{\partial x} = 0$$

$$\frac{\partial f'}{\partial y} = \frac{\partial f}{\partial y} + \lambda\frac{\partial g}{\partial y} = 0 \qquad \text{E3}$$

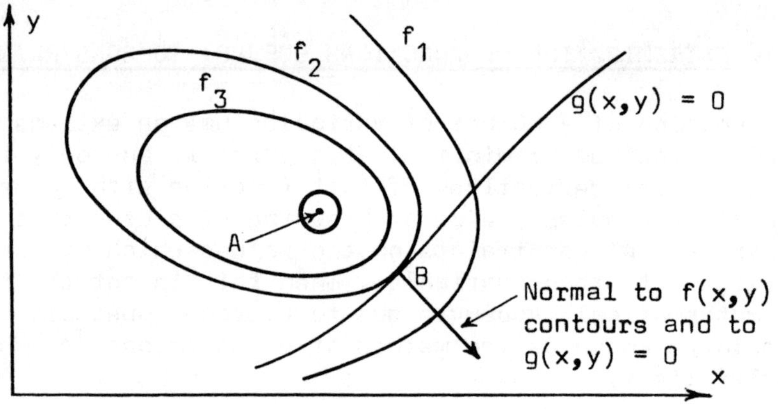

Fig (E1): Illustration of the concept of an extreme value of a function subject to a constraining relationship between the independent variables.

The essence of equation (E3) is that an extremisation problem subject to some given constraint has been converted to one of simple extremisation with no constraining relationship between the variables. The constant λ is known as an undetermined multiplier since it need never be evaluated explicitly. Should the function f be subject to n constraint expressions like g(x,y), then there will be n such multipliers involved. Thus, the equation (E2) will have its RHS augmented to involve n functions of the form $\lambda_i g_i$, $i = 1,\ldots,n$.

Example

Let it be required to determine the dimensions of the

rectangle of figure (E2) which is totally enclosed by the ellipse, such that the area of the rectangle is a maximum.

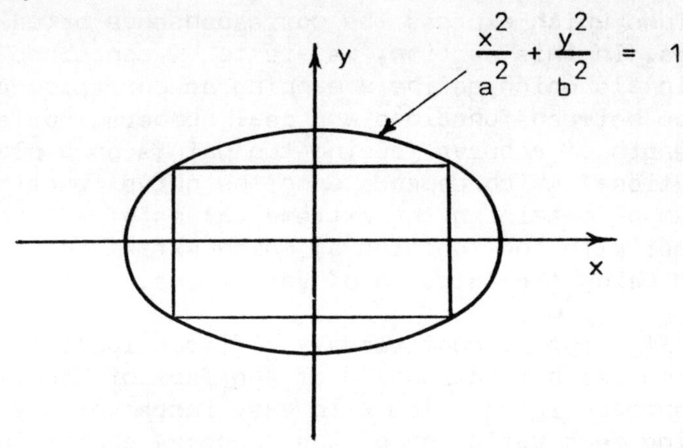

Fig (E2): Maximisation of the area of a rectangle imbedded within an ellipse.

The function to be maximised is the area A = xy subject to the constraining ellipse function. We therefore form the new function:

$$L = A + \lambda\left(\frac{x^2}{a^2} + \frac{y^2}{b^2} - 1\right)$$

On partial differentiation of L with respect to x and y and equating to zero, we obtain:

$$y + \frac{2\lambda x}{a^2} = 0 \quad \text{and} \quad x + \frac{2\lambda y}{b^2} = 0$$

On eliminating the Lagrange multiplier and on substitution into the equation of the ellipse, we obtain the co-ordinates of the corners of the rectangle which has maximum area.

$$x = \pm\frac{a}{\sqrt{2}} \quad : \quad y = \pm\frac{b}{\sqrt{2}}$$

E2. EXTREMISATION OF FUNCTIONALS SUBJECT TO CONSTRAINTS

Section (E1) was concerned with the extremisation of functions which express the correspondence between real numbers. In this section, we are to be concerned with functionals which define a mapping or correspondence, but now between functions and real numbers. For example, the length of a curve joining two points on a plane is a functional which depends upon the curve function. The problem of obtaining the extreme values of a functional together with the function(s) which extremise it can be solved using the Calculus of Variations.

Thus, if a set of continuously differentiable functions x extremises a functional J of the form of equation (E4) the approach is to allow x to vary incrementally (whilst ensuring zero variation of the boundary conditions) and to determine the total variation of J allowing a first-order Taylor's expansion about x and $\dot{x}$.

$$J \triangleq \int_a^b L(x,\dot{x},t)dt \qquad \text{E4}$$

On manipulating and setting the variation of J to zero (for an extremum), it is found that x satisfies the Euler equations:

$$\frac{\partial L(x,\dot{x},t)}{\partial x_i} - \frac{d}{dt}\frac{\partial L(x,\dot{x},t)}{\partial \dot{x}_i} = 0 : i = 1,..,n \qquad \text{E5}$$

The conditions of solution when the end point is variable is considered in standard texts (eg, 47). It is to be noted that equation (E5) is of identical form to Lagrange's equations where, with L representing the Lagrangian, a similar derivation has been employed.

When these ideas are considered for application to the determination of optimal controls for engineering systems we find some differences in problem statement. It is assumed that the plant may be described by the set of differential equations of expression (E6), where 'u' is the control vector.

$$\dot{x} = f(x,u,t) \qquad \text{E6}$$

It is also commonly required to determine that u which transfers the state vector x between two points whilst extremising the functional performance index of equation (E7).

$$J = \int_{t_i}^{t_f} L(x,u,t)dt \qquad \text{E7}$$

where t_i, t_f are the initial and final times of the manoeuvre.

It is possible to handle the control vector u by simple augmentation of the state vector, but the addition of the state equations (E6) imposes a constraint on x and u. However, the considerations of the previous section show that these can be handled by the use of undetermined multipliers. (Note that since we are now dealing with functionals, the multipliers themselves will be functions rather than scalars). One approach is to formulate a new cost function from J incorporating the undetermined multipliers and the constraints of the plant equations and to use the Euler equations to obtain the solution. Alternatively, one can follow the following method of solution.

First, let us generalise the cost function to include a terminal-condition cost function, which is found useful in certain problems.

$$J = S(x(t_f),t_f) + \int_{t_i}^{t_f} L(x,u,t)dt \qquad \text{E8}$$

Now form a new cost function including the plant and undetermined multiplier.

$$J' = S(x(t_f),t_f) + \int_{t_i}^{t_f} \{L(x,u,t) + \lambda^T(t)[f - \dot{x}]\}dt$$

For convenience, define the Hamiltonian H (often referred to as the state function of Pontryagin):

$$H(x,u,\lambda,t) \triangleq L(x,u,t) + \lambda^T f(x,u,t) \qquad \text{E9}$$

Now, on integrating the final part of the expression for J' by parts:

$$J' = S(x(t_f),t_f) - \lambda^T(t_f)x(t_f) + \lambda^T(t_i)x(t_i) + \int_{t_i}^{t_f} (H + \dot{\lambda}^T x)dt \qquad \text{E10}$$

Consider now the variation in J' due to variations in u at times t_i and t_f.

$$\delta J' = \left(\frac{\partial S}{\partial x} - \lambda^T\right)\delta x\Big|_{t_f} + \lambda^T\delta x\Big|_{t_i} + \int_{t_i}^{t_f} \left\{\left(\frac{\partial H}{\partial x} + \dot{\lambda}^T\right)\delta x + \frac{\partial H}{\partial u}\delta u\right\}dt \qquad \text{E11}$$

The objective is now to select $\lambda(t)$ such that the coefficients of δx vanish. This is accomplished by setting:

$$\dot{\lambda}^T = -\frac{\partial H}{\partial x}$$

subject to the boundary conditions:

$$\lambda^T(t_i) = 0 \quad \text{and} \quad \lambda^T(t_f) = \frac{\partial S}{\partial x}$$

Thus, for an extremum of J', we have the condition:

$$\frac{\partial H}{\partial u} = 0 \qquad \text{E12}$$

The procedure for the solution of the optimal control problem is therefore to formulate H and determine the optimal H_o using equation (E12) and hence the optimal control u_o and to solve equations (E13) subject to the above boundary conditions.

$$\dot{x} = \frac{\partial H_o}{\partial \lambda} \quad : \quad \dot{\lambda} = -\frac{\partial H_o}{\partial x} \qquad \text{E13}$$

There are two comments which are in order:

(1) Expression (E12) is a statement of extremisation if and only if u is completely unconstrained, ie not bounded or a member of a specified set. However, in what was to become known as the Extremum Principle, Pontryagin showed that irrespective of constraints on u, u_o must still be chosen to extremise H.

(2) The solution obtained for u_o is not a control law (ie derived from a knowledge of the state vector) and therefore is to be interpreted as the optimal open-loop control to be applied to achievo the objectives. It is applied in response to initial conditions rather than to an input.

For the purpose of closed-loop control we require an approach which yields a control law $u_o(x,t)$ and which eliminates the necessity of solving a two-point boundary value problem. With these stipulations, the computational problem becomes excessive except for the particular case of the linear system requiring the extremisation of a quadratic performance index. We therefore consider the system equation and cost function of equations (E14,15):

$$\dot{x} = Ax + Bu \qquad \text{E14}$$

$$J = \tfrac{1}{2}(x^T Sx)\Big|_{t_f} + \tfrac{1}{2}\int_{t_i}^{t_f} (x^T Px + u^T Qu)dt \qquad E15$$

The equations (E14,15) are now in state-variable format with T indicative of transposition and where the weighting matrices S,P,Q are positive definite. As already shown, it is necessary to solve equations (E12,13), thus:

$$H = \tfrac{1}{2}x^T Px + \tfrac{1}{2}u^T Qu + \lambda^T(Ax + Bu)$$

$$\dot{\lambda} = -Px - A^T\lambda \qquad E16$$

$$0 = Qu + B^T\lambda$$

We therefore see that the optimal control is given by:

$$u_o = -Q^{-1}B^T\lambda \qquad E17$$

Now the terminal condition for this linear-quadratic problem is obviously:

$$\lambda(t_f) = Sx\Big|_{t_f} \qquad E18$$

Further, as a consequence of the Principle of Optimality (that a process which is optimal from A to B is also optimal from C to B, where C is an intermediate point on the optimal trajectory), every section of the trajectory from t_i to t_f is itself an optimal trajectory for which the terminal condition of equation (E18) must apply. The only way for this to be true is if:

$$\lambda(t) = S(t)x(t) \qquad E19$$

If equation (E19) is substituted into (E18) we arrive at the matrix Riccati equation:

$$\dot{S} = -SA - A^{T}S + SBQ^{-1}B^{T}S - P \qquad \text{E20}$$

On solving for S, the optimal control law becomes:

$$u_{o} = -Q^{-1}B^{T}Sx \qquad \text{E21}$$

If the system is controllable and time-invariant, and if P and Q are constant, the matrix S itself is constant when the final time t_f is infinite. Equation (E20) is therefore equated to the null matrix and solved (74).

The interested reader is directed towards an alternative presentation of the optimal control problem by way of dynamic programming which is introduced in a lucid manner by McCausland (68).

References

1. MAYR, O.; 'The origins of feedback control'; MIT Press; 1970.

2. BELL, R. and de PENNINGTON, A.; 'Active compensation of lightly damped electrohydraulic cylinder drives using derivative signals'; Proc.I.Mech.Eng. Vol 184; 1969-70; pp 83-98.

3. TUSTIN, A., ALLANSON, J.T., LAYTON, M.M. and JAKEWAYS, R.; 'Design of systems for automatic control of the position of massive objects'; Proc. IEE; Vol 105, Part C; 1958.

4. TOWILL, D.R.; 'Coefficient plane design of instrument servos for shipboard use'; 3rd Ship Control Systems Symposium, Bath; 1972; Paper IV C-3.

5. TOWILL, D.R.; 'Optimum transfer functions for feedback control systems with plant input saturation'; Radio & Electronic Engr.; Vol 38; 1969; pp 233-47.

6. BOSLEY, M.J. and LEES, F.P.; 'A survey of simple transfer function derivations from high-order state variable models'; Automatica; Vol 8; 1972; pp 765-75.

7. TOWILL, D.R.; 'Coefficient plane models for control system analysis and design'; Research Studies Press; 1981.

8. NELDER, J.A. and MEAD, R.; 'A simplex method for function minimization'; Computer J.; Vol 5; 1965; pp 308-12.

9. ASHWORTH, M.J. and TOWILL, D.R.; 'Computer-aided design of tracking systems'; Radio & Electronic Engr.; Vol 48; 1978; pp 479-92.

10. TOWILL, D.R.; 'Low-order modelling: Tools or Toys' IEE Conf.; Computer-aided control system design; Cambridge, UK; 1973.

11. AXELBY, G.; 'Synthesis of feedback systems with a minimum lead for a specified performance'; Trans. IRE; 1956; pp 56-73.

12. BIERNSON, G.; 'Quick methods for evaluating the closed-loop poles of feedback control systems'; Trans. AIEE (Apps.& Ind.); Vol 72; 1953; pp 53-70.

13. KAN CHEN; 'A quick method for estimating closed-loop poles of control systems'; Trans. AIEE; Vol 76; 1957.

14. TOWILL, D.R. and MEHDI, Z.; 'Prediction of the transient response sensitivity of high-order linear systems using low-order models'; J.Meas.& Control; Vol 3; 1970; pp T1-9.

15. MATSUBARA, M.; 'On the equivalent dead-time'; Trans. IEEE; Vol AC-10; 1965; pp 464-66.

16. CHEN, C.F. and SHIEH, L.S.; 'An algebraic method for control system design'; Int.J.Control; Vol 11; 1970; pp 717-39.

17. SHAMASH, Y.; 'Linear system reduction using Pade approximation to allow retention of dominant modes' Int.J.Control; Vol 21; 1975.

18. CHUANG, S.C.; 'A method for linear systems order reduction'; IEE Conf.; Computer-aided Design; Cambridge, UK; 1973; pp 213-19.

19. WHALLEY, R.; 'Transfer function reduction'; Proc. I.Mech.E.; Vol 190; 1976; pp 643-51.

20. LEVY, E.C.; 'Complex curve fitting'; Trans.IRE; Vol AC-4; 1959; pp 37-43.

21. SANANTHANAN, C.K. and KOERNER, J.; 'Transfer function synthesis as the ratio of two complex

polynomials'; Trans.IEEE; Vol AC-8; 1963; pp 56-8.

22. PAYNE, P.A.; 'An improved technique for transfer function synthesis from frequency response data'; Trans.IEEE; Vol AC-15; 1970; pp 480-83.

23. PAYNE, P.A., TOWILL, D.R. and BAKER, K.J.; 'Predicting servomechanism dynamic performance variation from limited production test data'; Radio & Electronic Engr.; Vol 40; 1970; pp 275-88.

24. LATHROP, R.C. and GRAHAM, D.C.; 'The transient performance of servomechanisms with derivative and integral control'; Applications & Industry, No.11; Mar 1954; pp 10-17.

25. TOWILL, D.R.; 'Analysis and synthesis of feedback compensated third-order control systems via the coefficient plane'; Radio & Electronic Engr.; Vol 32; 1966; pp 119-31.

26. TOWILL, D.R.; 'Coefficient plane synthesis of zero velocity-lag servomechanisms'; Ibid; Vol 34; 1967; pp 323-34.

27. ASHWORTH, M.J.; 'Computer-aided design of tracking systems'; M.Phil Thesis; RNEC, Plymouth, UK; 1975.

28. ASHWORTH, M.J. and TOWILL, D.R.; 'Coefficient plane analysis, synthesis and testing of servo systems'; 4th Pittsburgh Conf.; Modelling and Simulation; University of Pittsburgh, USA; 1976.

29. BOX, G.E. and JENKINS, G.M.; 'Time series analysis: forecasting and control'; Holden-Day; 1970.

30. GIBILARO, L.G. and LEES, F.P.; 'The reduction of complex transfer function models to simple models using the method of moments'; Chem.Eng.Sci.; Vol 24; 1969; pp 85-93.

31. BODE, H.W.; 'Network analysis and feedback amplifier design'; Van Nostrand; 1945; p 285.

32. HOROWITZ, I.M.; 'Synthesis of feedback systems'; Academic Press; 1963; pp 280-84.

33. LINDORFF, D.P.; 'Theory of sampled data systems'; Wiley; 1965; pp 152-56.

34. LINDORFF, D.P.; 'Sensitivity in sampled-data systems'; Trans.IEEE; Vol AC-8; 1963; pp 120-25.

35. SIDI, M.; 'Synthesis of sampled feedback systems for prescribed time domain tolerances'; Int.J. Control; Vol 26, No.3; 1977; pp 445-61.

36. SIDI, M.; 'On maximization of gain-bandwidth in sampled-data systems'; Int.J.Control; Vol 32, No.6 1980; pp 1099-1109.

37. KLAFIN, J.F. and KRISHNAN, V.; 'Linear sensitivity analysis applied to a two-loop system with feedback variations'; Int.J.Control; Vol 15; 1972; pp 305-17.

38. HOROWITZ, I.M. and SHAKED, U.; 'Superiority of transfer function over state-variable methods in linear time-invariant feedback system design'; Trans.IEEE; Vol AC-20; 1975; pp 84-97.

39. PORTER, B. and CROSSLEY, R.; 'Modal control theory and applications'; Taylor & Francis; 1972.

40. TOWILL, D.R. and MEHDI, Z.; 'A new approach to system transient response sensitivity'; Int.J. Control; Vol 15, No.2; 1972; pp 319-31.

41. LIOU, M.L.; 'A novel method of evaluating transient response'; Proc.IEEE; Vol 54; 1966; pp 20-23.

42. TOWILL, D.R. and PAYNE, P.A.; 'Frequency domain approach to automatic testing of control systems'; Radio & Electronic Engr.; Vol 41; 1971; pp 51-60.

43. CHUBB, B.A.; 'Modern analytical design of instrument servomechanisms'; Addison-Wesley; 1971; pp 121-43.

44. LEY, B.J.; 'Computer-aided analysis and design for electrical engineers'; Holt, Rinehart & Winston; 1970; pp 522-95.

45. FORTMANN, T.E. and WILLIAMSON, D.; 'Design of low-order observers for linear feedback control laws'; Trans.IEEE; Vol AC-17; 1972; pp 301-307.

46. KORTUM, W.; 'Computational techniques in optimal state estimation - a tutorial review'; J.ASME; (Dynamic Systems, Measurement & Control); Vol 101; 1979; pp 99-107.

47. SCHULTZ, D.G. and MELSA, J.L.; 'State functions and linear control systems'; McGraw-Hill; 1967; pp 365-422.

48. HOROWITZ, I.M. and SIDI, M.; 'Synthesis of feedback systems with large plant ignorance for prescribed time-domain tolerances'; Trans.IEEE; Vol 20; 1972; pp 84-97.

49. ASHWORTH, M.J. and TOWILL, D.R.; 'Realization of plant sensitivity specifications via multi-loop feedback compensation'; IFAC Symposium: Computer-Aided Design of Control Systems; Zurich; 1979.

50. HOROWITZ, I.M.; 'Optimum loop transfer function in single-loop minimum-phase feedback systems'; Int. J.Control; Vol 18; 1973; pp 97-113.

51. HOROWITZ, I.M. and SIDI, M.; 'Optimum synthesis of non-minimum phase feedback systems with plant uncertainty'; Int.J.Control; Vol 27; 1978; pp 361-386.

52. GERA, A. and HOROWITZ, I.M.; 'Optimization of the loop transfer function'; Int.J.Control; Vol 31; 1980; pp 389-98.

53. ASHWORTH, M.J. and TOWILL, D.R.; 'An assessment of the potential of sensitivity design in the optimization of control strategies for manoeuvring vessels'; LAN Symposium: Ship Steering & Automatic

Control; Genoa, Italy; 1980; pp 245-58.

54. SIDI, M.; 'Synthesis of feedback systems with large plant ignorance for prescribed time-domain tolerances'; PhD Thesis; Weizmann Institute of Science, Rehovot, Israel; 1973.

55. KRISHNAN, K.R. and CRUIKSHANKS, A.; 'Frequency domain design of feedback systems for specified insensitivity of time-domain response to parameter variation'; Int.J.Control; Vol 25; 1977; pp 609-20.

56. ASHWORTH, M.J. and TOWILL, D.R.; 'CACSD for sensitivity reduction via coefficient-plane modelling'; IEE conf.; Control and its Applications; University of Warwick, UK; 1981.

57. HOROWITZ, I.M. and SIDI, M.; 'Synthesis of cascaded multiple-loop feedback systems with large plant parameter ignorance'; Automatica; Vol 9; 1973; pp 589-600.

58. ABKOWITZ, M.A.; 'Lectures on ship hydrodynamics: steering and manoeuvrability'; Report No. Hy-5; Hydro-og Aerodynamisk Lab., Lyngby, Denmark; 1964.

59. CHISLETT, M.S. and STROM-TEJSEN, J.; 'Planar-motion mechanism tests and full scale steering and manoeuvring predictions for a Mariner-Class vessel' Ibid; Report No. Hy-6; 1965.

60. ASHWORTH, M.J.; 'Determination of a toleranced linear model of a ship'; RNEC, Plymouth, UK; CSR No. 64; 1979.

61. CLARKE, D.W. and GAWTHROP, P.J.; 'Self-tuning controller'; Proc.IEE; Vol 122; 1975; pp 929-34.

62. HOROWITZ, I.M.; 'Improvement in quantitative non-linear feedback design by cancellation'; Int.J. Control; Vol 34; 1981; pp 547-60.

63. HOROWITZ, I.M., GOLUBEV, B. and KOPELMAN, T.; 'Flight control design based on nonlinear model

with uncertain parameters'; J.Guidance & Control; Vol 3, No.2; 1980; pp 113-18.

64. MACFARLANE, A.G.J.; 'A survey of some recent results in linear multivariable feedback theory'; Automatica; Vol 8; 1972; pp 455-85.

65. HOROWITZ, I.M.; 'Quantitative synthesis of uncertain multiple input-output feedback systems'; Int. J.Control; Vol 30; 1979; pp 81-106.

66. NEWTON, G.C., GOULD, L.A. and KAISER, J.F.; 'Analytical design of linear feedback controls'; Wiley; 1967; pp 371-81.

67. NEWLAND, D.E.; 'Random vibrations and spectral analysis'; Longman; 1975; Chapter 5.

68. McCAUSLAND, I.; 'Introduction to optimal control'; Wiley; 1969; pp 31-39.

69. GARNELL, P. and EAST, D.J.; 'Guided weapon control systems'; Pergamon Press; 1977; Chapter 10.

70. WIENER, N.; 'Extrapolation, interpolation and smoothing of stationary time series'; Wiley and Technology Press of MIT; New York; 1949.

71. BODE, H.W. and SHANNON, C.E.; 'A simplified derivation of linear least-square smoothing and prediction theory'; Proc.IRE; Vol 38; 1950; pp 417-25.

72. WHALLEY, R.; 'The control of marine propulsion plant'; PhD Thesis; RNEC, Plymouth, UK; 1976.

73. LANING, J.H. and BATTIN, R.H.; 'Random processes in automatic control'; McGraw-Hill; 1956.

74. ATHANS, M. and FALB, P.L.; 'Optimal control: an introduction to the theory and its applications'; McGraw-Hill; 1966.

75. ASHWORTH, M.J. and TOWILL, D.R.; 'Autopilot design based on nonlinear ship models with uncertain

parameters'; 6th Ship Control Systems Symposium, Ottawa, Canada; Oct 1981.

76. ASTROM, K.J. and EYKHOFF, P.; 'System indentification: a survey'; Automatica; Vol 7; 1971; pp 123-162.

77. WELLSTEAD, P.E., EDMUNDS, J.M., PRAGER, D.L. and ZANKER, P.M.; 'Pole-zero assignment self-tuning regulators'; Int.J.Control; Vol 30; 1979; pp 1-26.

78. KALLSTROM, C.G., ASTROM, K.J., THORELL, N.E., ERIKSSON, J. and STEN, L.; 'Adaptive autopilots for tankers'; Automatica; Vol 15; 1979; pp 241-54.

79. GRAHAM, D. and LATHROP, R.C.; 'The synthesis of optimum transient responses: criteria and standard forms'; Trans.AIEE(Applications & Industry); Vol 72; 1953; p 273.

80. BLACKBURN, T.R. and VAUGHAN, D.R.; 'Applications of linear optimal control and filtering theory to the Saturn V launch vehicle'; Trans.IEEE; Vol AC-16; 1971; pp 799-806.

81. BROWN, J.M.; 'A hybrid classical-modern control theory approach to optimal design'; IEE CACSD Conference; Cambridge, UK; 1973; pp 55-62.

82. TOWILL, D.R.; 'Coefficient plane models for tracking system design'; Radio & Electronic Engr.; Vol 48, No.10; 1975; pp 465-71.

83. CHANG, S.S.L.; 'Synthesis of optimum control systems'; McGraw-Hill; 1961.

84. CHEN, C.T.; 'Analysis and synthesis of linear control systems'; Holt, Rinehart & Winston; 1975; pp 202-219.

85. LEAKE, R.J.; 'Return difference Bode diagram for optimal system design'; Trans.IEEE; Vol AC-10, No.3; 1965; pp 342-4.

86. KALMAN, R.E.; 'When is a linear system optimal?'; J.Basic Eng.; Ser.D, Vol 86; 1964; pp 51-60.

87. ROSENBROCK, H.H. and McMORRAN, P.D.; 'Good, bad or optimal'; Trans.IEEE; Vol AC-16; 1971; pp 552-3.

88. RIDDLE, A.C. and ANDERSON, B.D.; 'Spectral factorization: computational aspects'; Trans. IEEE Vol AC-11; 1966; pp 764-5.

89. LUENBERGER, D.G.; 'Observing the state of a linear system'; Trans.IEEE; MIL-8; 1964; p 74.

90. FORTMANN, T.E.; 'Design of low-order observers for linear feedback control laws'; Trans.IEEE; Vol AC-17, No.3; 1972; pp 301-7.

91. LEONDES, C.T. and NOVAK, L.M.; 'Optimal minimal-order observers for discrete-time systems: a unified theory'; Automatica; Vol 8; 1972;pp379-87.

Index

Accuracy, dynamic 23
Algorithms,
  modelling, 189
  boundary circles, 195
Aliassing, 66
Anchor zeros, 88
Autocorrelation, 129

Bandwidth, 12,25,26
  effect on plant
  saturation, 132
Bode,H.W. 60,139,199
Bode diagram,
  synthesis of the
  optimal function, 181
Bode's ideal loop
  function, 207,210
Bode's theorems, 200
Boundary,
  circles, 195
  of loop function, 102
Butterworth,
  functions, 48,177,185

Chang,S.S.L., 163
Characteristics,
  ideal loop function 207
  system functions, 199
Chen,C.F., 40
Chuang,S.C., 43
Closed-loop,
  configurations, 4
  tolerances, 105
Coefficient plane 49
Component tolerancing 75
CAD 111
Continued fraction
  model 40
Control effort,
  constraint of, 157
Cost functions, 158-62
Cost of feedback 63
Curve fitting, 43

Degrees of
  freedom, 4,10,61
Disturbance,
  control via Nichols
  chart, 152
  external noise, 141
  feedforward, 8
  rejection, 7
Dominant,
  system function, 27
Dynamic
  accuracy, 23
  achievement of
  response, 12
  specifications, 24-7
  uncertainty, 118

Effective time
  constants, 36
Equivalent time delay 38
Error coefficients 25
Estimation of noise
  levels, 128
Extremization,

of functions, 223
of functionals, 226

Far-off poles, 36,94
Feedback,
benefits of, 5
cost of, 65
effect on non-linearities, 20
equality of positive and negative areas, 202
minor-loop, 17
noise, 64,93,146
objectives, 1
observer, 179
reduction of system sensitivity, 5
Frequency response,
approach to sensitivity, 69
of sampled signals, 65
specifications, 26
synthesis of the loop function, 97

Hold, zero-order 65
Horowitz,I.M., 94,97,116 214

Ideal loop,
alternatives, 214
characteristic 210
Ignorance,
of the system and its environment, 56
Integral control, 10
Integral,
of absolute error, 159
of error squared, 159
ITAE criterion, 12,48,159

Kalman's equation, 181
Kan-Chen,
dominant model, 34

Lagrange,
undetermined multipliers, 224
Leake,
optimal function, 181
Least-squares fit, 43,189
Levy,E.C., 43
Loop characteristics,
Bodes, 210
alternatives, 214
Loop function,
optimization, 114
stability, 105
synthesis, 81,97
via Nichols chart, 108
Loop magnitude,
boundaries, 98,195
computation, 113
Low-order modelling, 31
reasons for, 32

Matrix-Ricatti,
equation, 231
Matsubara,m., 38
Mean-square error,
via simulation, 135
Modelling,
continued fraction, 40
from frequency response, 43
from loop frequency response, 34
from system transfer function, 38
from time response, 51
loss of information, 33
low-order, 31
reasons for, 32
standard forms, 47
summary, 52
Moments,
method of, 51
Monte-Carlo,
simulation, 78,111

Nichols chart,
loop synthesis, 108
Noise,
estimation of MSE levels, 128
feedback, 64,93,146
optimization of parameters, 132
transmission, 93
Noise disturbance, 127
effects of, 128
external, 141
with variation of the plant, 141
Nonlinear systems, 57
behaviour, 18
compensation, 19,120
controllers, 120
design, 118
Non-minimum,
singularities, 12,18,45 67

Normalization,
frequency, 48

Observer feedback, 179
Optimal control,
summary, 187
Optimality, 223
Bode diagram, 181
indices of, 158
root-square locus, 170
solution at high gain levels, 176
Transfer function approach, 162
Optimum systems,
wrt step input, 167
wrt ramp input, 167
Optimization,
in presence of noise, 132
of loop function, 114
Overshoot, peak 25

Parametric variation,
causes of, 58
with disturbance, 141
Parseval's theorem 162,219
Performance,
discrete index, 160
indices, 158
specifications, 23
Phase,
non-minimum, 12,42,67
Plants,
ignorance, 56
ill conditioned, 12
input excitation, 27
unstable, 18
variation, 55,58
Poles,
sensitivity of, 71
Pole-zero synthesis,
loop function, 88
Power density,
spectrum, 129

Reactance-integral,
theorem, 202
Resistance-Integral,
theorem, 200
Resonance, 12,14
peak, 26
Rise time 24
Root-locus,
synthesis, 82
Root-square locus,
synthesis, 170

Sampling,
effects on system sensitivity, 65
maximum bandwidth, 68
selection of period 68
Sensitivity, 55
analysis, 68
cost of reduction, 63
frequency response, 69

function, 59
pole-zero migration, 71
reduction, 5,88
transient response, 72
Settling time, 24
Ship dynamics, 20,118
Sidi,M., 107,152
Simulation,
estimation of MSE, 135
Specifications,
closed-loop, 105
frequency domain, 26
time domain, 24
Spectral factorization 166
Spectrum,
power density, 129
Stability, 12-13,105
Standard forms, 47
Sufficiency theorem, 204
Synthesis,
graphical, loop, 97
with far-off poles, 94
with variable plant zeros, 96
zero assignment, 84

Task definition, 23
Time delay,
equivalent, 38
Time domain,
specifications, 24
Tolerancing, 75
Towill,D.R., 48,77,132
Transfer function,
approach to the optimum system, 162
optimal wrt step and ramp inputs, 167
Transient response,
sensitivity function 72

Variation,
parametric, 58

Weighted least-squares 43
Whalley,R., 43
Wiener filter, 137
W-plane, 67

Zero-assignment, 84
Zeros, variable, 96
Z-plane, 67